Screw Compressors

N. Stosic I. Smith A. Kovacevic

Screw Compressors

Mathematical Modelling and Performance Calculation

With 99 Figures

Springer

Prof. Nikola Stosic
Prof. Ian K. Smith
Dr. Ahmed Kovacevic

City University
School of Engineering and Mathematical Sciences
Northampton Square
London
EC1V 0HB
U.K.
e-mail: n.stosic@city.ac.uk
i.k.smith@city.ac.uk
a.kovacevic@city.ac.uk

ISBN-13 978-3-642-06350-3 e-ISBN-13 978-3-540-26228-2

Springer is a part of Springer Science+Business Media
springeronline.com

Softcover reprint of the hardcover 1st edition 2005

Cover design: medio, Berlin

Preface

Although the principles of operation of helical screw machines, as compressors or expanders, have been well known for more than 100 years, it is only during the past 30 years that these machines have become widely used. The main reasons for the long period before they were adopted were their relatively poor efficiency and the high cost of manufacturing their rotors. Two main developments led to a solution to these difficulties. The first of these was the introduction of the asymmetric rotor profile in 1973. This reduced the blow-hole area, which was the main source of internal leakage by approximately 90%, and thereby raised the thermodynamic efficiency of these machines, to roughly the same level as that of traditional reciprocating compressors. The second was the introduction of precise thread milling machine tools at approximately the same time. This made it possible to manufacture items of complex shape, such as the rotors, both accurately and cheaply.

From then on, as a result of their ever improving efficiencies, high reliability and compact form, screw compressors have taken an increasing share of the compressor market, especially in the fields of compressed air production, and refrigeration and air conditioning, and today, a substantial proportion of compressors manufactured for industry are of this type.

Despite, the now wide usage of screw compressors and the publication of many scientific papers on their development, only a handful of textbooks have been published to date, which give a rigorous exposition of the principles of their operation and none of these are in English.

The publication of this volume coincides with the tenth anniversary of the establishment of the Centre for Positive Displacement Compressor Technology at City University, London, where much, if not all, of the material it contains was developed. Its aim is to give an up to date summary of the state of the art. Its availability in a single volume should then help engineers in industry to replace design procedures based on the simple assumptions of the compression of a fixed mass of ideal gas, by more up to date methods. These are based on computer models, which simulate real compression and expansion processes more reliably, by allowing for leakage, inlet and outlet flow and other losses, and the assumption of real fluid properties in the working process. Also, methods are given for developing rotor profiles, based on the mathematical theory of gearing, rather than empirical curve fitting. In addition, some description is included of procedures for the three dimensional modelling of heat and fluid flow through these machines and how interaction between the rotors and the casing produces performance changes, which hitherto could not be calculated. It is shown that only a relatively small number of input parameters is required to describe both the geometry and performance of screw compressors. This makes it easy to control the design process so that modifications can be cross referenced through design software programs, thus saving both computer resources and design time, when compared with traditional design procedures.

All the analytical procedures described, have been tried and proven on machines currently in industrial production and have led to improvements in performance and reductions in size and cost, which were hardly considered possible ten years ago. Moreover, in all cases where these were applied, the improved accuracy of the analytical models has led to close agreement between predicted and measured performance which greatly reduced development time and cost. Additionally, the better understanding of the principles of operation brought about by such studies has led to an extension of the areas of application of screw compressors and expanders.

It is hoped that this work will stimulate further interest in an area, where, though much progress has been made, significant advances are still possible.

London,
February 2005

Nikola Stosic
Ian Smith
Ahmed Kovacevic

Notation

A	Area of passage cross section, oil droplet total surface
a	Speed of sound
C	Rotor centre distance, specific heat capacity, turbulence model constants
d	Oil droplet Sauter mean diameter
e	Internal energy
$\mathbf{f}$	Body force
h	Specific enthalpy $h = h(\theta)$, convective heat transfer coefficient between oil and gas
$\mathbf{i}$	Unit vector
$\mathbf{I}$	Unit tensor
k	Conductivity, kinetic energy of turbulence, time constant
m	Mass
$\dot{m}$	Inlet or exit mass flow rate $\dot{m} = \dot{m}(\theta)$
p	Rotor lead, pressure in the working chamber $p = p(\theta)$
P	Production of kinetic energy of turbulence
$\mathbf{q}$	Source term
$\dot{Q}$	Heat transfer rate between the fluid and the compressor surroundings $\dot{Q} = \dot{Q}(\theta)$
r	Rotor radius
s	Distance between the pole and rotor contact points, control volume surface
t	Time
T	Torque, Temperature
$\mathbf{u}$	Displacement of solid
U	Internal energy
W	Work output
$\mathbf{v}$	Velocity
$\mathbf{w}$	Fluid velocity
V	Local volume of the compressor working chamber $V = V(\theta)$
$\dot{V}$	Volume flow

x Rotor coordinate, dryness fraction, spatial coordinate
y Rotor coordinate
z Axial coordinate

Greek Letters

α Temperature dilatation coefficient
Γ Diffusion coefficient
ε Dissipation of kinetic energy of turbulence
η_i Adiabatic efficiency
η_t Isothermal efficiency
η_v Volumetric efficiency
φ Specific variable
ϕ Variable
λ Lame coefficient
μ Viscosity
ρ Density
σ Prandtl number
θ Rotor angle of rotation
ζ Compound, local and point resistance coefficient
ω Angular speed of rotation

Prefixes

d differential
Δ Increment

Subscripts

eff Effective
g Gas
in Inflow
f Saturated liquid
g Saturated vapour
ind Indicator
l Leakage
oil Oil
out Outflow
p Previous step in iterative calculation
s Solid
T Turbulent
w pitch circle
1 main rotor, upstream condition
2 gate rotor, downstream condition

Contents

1 Introduction

The screw compressor is one of the most common types of machine used to compress gases. Its construction is simple in that it essentially comprises only a pair of meshing rotors, with helical grooves machined in them, contained in a casing, which fits closely round them. The rotors and casing are separated by very small clearances. The rotors are driven by an external motor and mesh like gears in such a manner that, as they rotate, the space formed between them and the casing is reduced progressively. Thus, any gas trapped in this case is compressed. The geometry of such machines is complex and the flow of the gas being compressed within them occurs in three stages. Firstly, gas enters between the lobes, through an inlet port at one end of the casing during the start of rotation. As rotation continues, the space between the rotors no longer lines up with the inlet port and the gas is trapped and thus compressed. Finally, after further rotation, the opposite ends of the rotors pass a second port at the other end of the casing, through which the gas is discharged. The whole process is repeated between successive pairs of lobes to create a continuous but pulsating flow of gas from low to high pressure.

These machines are mainly used for the supply of compressed air in the building industry, the food, process and pharmaceutical industries and, where required, in the metallurgical industry and for pneumatic transport. They are also used extensively for compression of refrigerants in refrigeration and air conditioning systems and of hydrocarbon gases in the chemical industry. Their relatively rapid acceptance over the past thirty years is due to their relatively high rotational speeds compared to other types of positive displacement machine, which makes them compact, their ability to maintain high efficiencies over a wide range of operating pressures and flow rates and their long service life and high reliability. Consequently, they constitute a substantial percentage of all positive displacement compressors now sold and currently in operation.

The main reasons for this success are the development of novel rotor profiles, which have drastically reduced internal leakage, and advanced machine tools, which can manufacture the most complex shapes to tolerances of the order of 3 micrometers at an acceptable cost. Rotor profile enhancement is still the most promising means of further improving screw compressors and rational procedures are now being developed both to replace earlier empirically derived shapes and also to vary the proportions of the selected profile to obtain the best result for the application for which the compressor is required. Despite their wide usage, due to the complexity of their internal geometry and the non-steady nature of the processes within them, up till recently, only approximate analytical methods have been available to predict their performance. Thus, although it is known that their elements are distorted both by the heavy loads imposed by pressure induced forces and through temperature changes within them, no methods were available to predict the magnitude of these distortions accurately, nor how they affect the overall performance of the machine. In addition, improved modelling of flow patterns within the machine can lead to better porting design. Also, more accurate determination of bearing loads and how they fluctuate enable better choices of bearings to be made. Finally, if rotor and casing distortion, as a result of temperature and pressure changes within the compressor, can be estimated reliably, machining procedures can be devised to minimise their adverse effects.

Screw machines operate on a variety of working fluids, which may be gases, dry vapour or multi-phase mixtures with phase changes taking place within the machine. They may involve oil flooding, or other fluids injected during the compression or expansion process, or be without any form of internal lubrication. Their geometry may vary depending on the number of lobes in each rotor, the basic rotor profile and the relative proportions of each rotor lobe segment. It follows that there is no universal configuration which would be the best for all applications. Hence, detailed thermodynamic analysis of the compression process and evaluation of the influence of the various design parameters on performance is more important to obtain the best results from these machines than from other types which could be used for the same application. A set of well defined criteria governed by an optimisation procedure is therefore a prerequisite for achieving the best design for each application. Such guidelines are also essential for the further improvement of existing screw machine designs and broadening their range of uses. Fleming et al., 1998 gives a good contemporary review of screw compressor modelling, design and application.

A mathematical model of the thermodynamic and fluid flow processes within positive displacement machines, which is valid for both the screw compressor and expander modes of operation, is presented in this Monograph. It includes the use of the equations of conservation of mass, momentum and energy applied to an instantaneous control volume of trapped fluid within the machine with allowance for fluid leakage, oil or other fluid injection, heat transfer and the assumption of real fluid properties. By simultaneous solution of these equations, pressure-volume diagrams may be derived of the entire admission, discharge and compression or expansion process within the machine.

A screw machine volume is defined by the rotor profile which is here generated by use of a general gearing algorithm and the port shape and size. This algorithm demonstrates the meshing condition which, when solved explicitly, enables a variety of rotor primary arcs to be defined either analytically or by discrete point curves. Its use greatly simplifies the design since only primary arcs need to be specified and these can be located on either the main or gate rotor or even on any other rotor including a rack, which is a rotor of infinite radius. The most efficient profiles have been obtained from a combined rotor-rack generation procedure.

The rotor profile generation processor, thermofluid solver and optimizer, together with pre-processing facilities for the input data and graphical post processing and CAD interface, have been incorporated into a design tool in the form of a general computer code which provides a suitable tool for analysis and optimization of the lobe profiles and other geometrical and physical parameters. The Monograph outlines the adopted rationale and method of modelling, compares the shapes of the new and conventional profiles and illustrates potential improvements achieved with the new design when applied to dry and oil-flooded air compressors as well as to refrigeration screw compressors.

The first part of the Monograph gives a review of recent developments in screw compressors.

The second part presents the method of mathematical definition of the general case of screw machine rotors and describes the details of lobe shape specification. It focuses on a new lobe profile of a slender shape with thinner lobes in the main rotor, which yields a larger cross-sectional area and shorter sealing lines resulting in higher delivery rates for the same tip speed.

The third part describes a model of the thermodynamics of the compression-expansion processes, discusses some modelling issues and compares the shapes of new and conventional profiles. It illustrates the potential improvements achievable with the new design applied to dry and oil-flooded air compressors as well as to refrigeration screw compressors. The selection of the best gate rotor tip radius is given as an example of how mathematical modelling may be used to optimise the design and the machine's operating conditions.

The fourth part describes the design of a high efficiency screw compressor with new rotor profiles. A well proven mathematical model of the compression process within positive displacement machines was used to determine the optimum rotor size and speed, the volume ratio and the oil injection position and jet diameter. In addition, modern design concepts such as an open suction port and early exposure of the discharge port were included, together with improved bearing and seal specification, to maximise the compressor efficiency. The prototypes were tested and compared with the best compressors currently on the market. The measured specific power input appeared to be lower than any published values for other equivalent compressors currently manufactured. Both the predicted advantages of the new rotor profile and the superiority of the design procedure were thereby confirmed.

1.1 Basic Concepts

Thermodynamic machines for the compression and expansion of gases and vapours are the key components of the vast majority of power generation and refrigeration systems and essential for the production of compressed air and gases needed by industry. Such machines can be broadly classified by their mode of operation as either turbomachines or those of the positive displacement type.

Turbomachines effect pressure changes mainly by dynamic effects, related to the change of momentum imparted to the fluids passing through them. These are associated with the steady flow of fluids at high velocities and hence these machines are compact and best suited for relatively large mass flow rates. Thus compressors and turbines of this type are mainly used in the power generation industry, where, as a result of huge investment in research and development programmes, they are designed and built to attain thermodynamic efficiencies of more than 90% in large scale power production plant. However, the production rate of machines of this type is relatively small and worldwide, is only of the order of some tens of thousands of units per annum.

Positive displacement machines effect pressure changes by admitting a fixed mass of fluid into a working chamber where it is confined and then compressed or expanded and, from which it is finally discharged. Such machines must operate more or less intermittently. Such intermittent operation is relatively slow and hence these machines are comparatively large. They are therefore better suited for smaller mass flow rates and power inputs and outputs. A number of types of machine operate on this principle such as reciprocating, vane, scroll and rotary piston machines.

In general, positive displacement machines have a wide range of application, particularly in the fields of refrigeration and compressed air production and their total world production rate is in excess of 200 million units per annum. Paradoxically, but possibly because these machines are produced by comparatively small companies with limited resources, relatively little is spent on research and development programmes on them and there are very few academic institutions in the world which are actively promoting their improvement.

One of the most successful positive displacement machines currently in use is the screw or twin screw compressor. Its principle of operation, as indicated in Fig. 1.1, is based on volumetric changes in three dimensions rather than two. As shown, it consists, essentially, of a pair of meshing helical lobed rotors, contained in a casing. The spaces formed between the lobes on each rotor form a series of working chambers in which gas or vapour is contained. Beginning at the top and in front of the rotors, shown in the light shaded portion of Fig. 1.1a, there is a starting point for each chamber where the trapped volume is initially zero. As rotation proceeds in the direction of the arrows, the volume of that chamber then increases as the line of contact between the rotor with convex lobes, known as the main rotor, and the adjacent lobe of the gate rotor

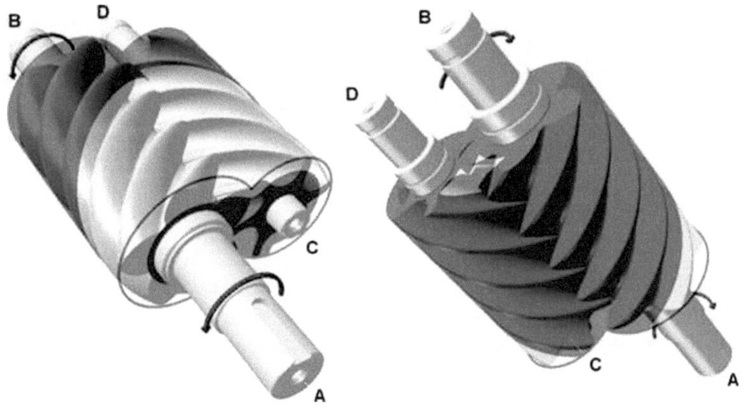

(a) View from Front and Top (b) View from Bottom and Rear

Fig. 1.1. Screw Compressor Rotors

advances along the axis of the rotors towards the rear. On completion of one revolution i.e. 360° by the main rotor, the volume of the chamber is then a maximum and extends in helical form along virtually the entire length of the rotor. Further rotation then leads to reengagement of the main lobe with the succeeding gate lobe by a line of contact starting at the bottom and front of the rotors and advancing to the rear, as shown in the dark shaded portions in Fig. 1.1b. Thus, the trapped volume starts to decrease. On completion of a further 360° of rotation by the main rotor, the trapped volume returns to zero.

The dark shaded portions in Fig. 1.1 show the enclosed region where the rotors are surrounded by the casing, which fits closely round them, while the light shaded areas show the regions of the rotors, which are exposed to external pressure. Thus the large light shaded area in Fig. 1.1a corresponds to the low pressure port while the small light shaded region between shaft ends B and D in Fig. 1.1b corresponds to the high pressure port.

Exposure of the space between the rotor lobes to the suction port, as their front ends pass across it, allows the gas to fill the passages formed between them and the casing until the trapped volume is a maximum. Further rotation then leads to cut off of the chamber from the port and progressive reduction in the trapped volume. This leads to axial and bending forces on the rotors and also to contact forces between the rotor lobes. The compression process continues until the required pressure is reached when the rear ends of the passages are exposed to the discharge port through which the gas flows out at approximately constant pressure.

It can be appreciated from examination of Fig. 1.1, is that if the direction of rotation of the rotors is reversed, then gas will flow into the machine through the high pressure port and out through the low pressure port and it will act as an expander. The machine will also work as an expander when rotating in the same direction as a compressor provided that the suction and discharge ports are positioned on the opposite sides of the casing to those shown since this is effectively the same as reversing the direction of rotation relative to the ports. When operating as a compressor, mechanical power must be supplied to shaft A to rotate the machine. When acting as an expander, it will rotate automatically and power generated within it will be supplied externally through shaft A.

The meshing action of the lobes, as they rotate, is the same as that of helical gears but, in addition, their shape must be such that at any contact position, a sealing line is formed between the rotors and between the rotors and the casing in order to prevent internal leakage between successive trapped passages. A further requirement is that the passages between the lobes should be as large as possible, in order to maximise the fluid displacement per revolution. Also, the contact forces between the rotors should be low in order to minimise internal friction losses. A typical screw rotor profile is shown in Fig. 1.2, where a configuration of 5–6 lobes on the main and gate rotors is presented. The meshing rotors are shown with their sealing lines, for the axial plane on the left and for the cross-sectional plane in the centre. Also, the clearance distribution between the two rotor racks in the transverse plane, scaled 50 times (6) is given above.

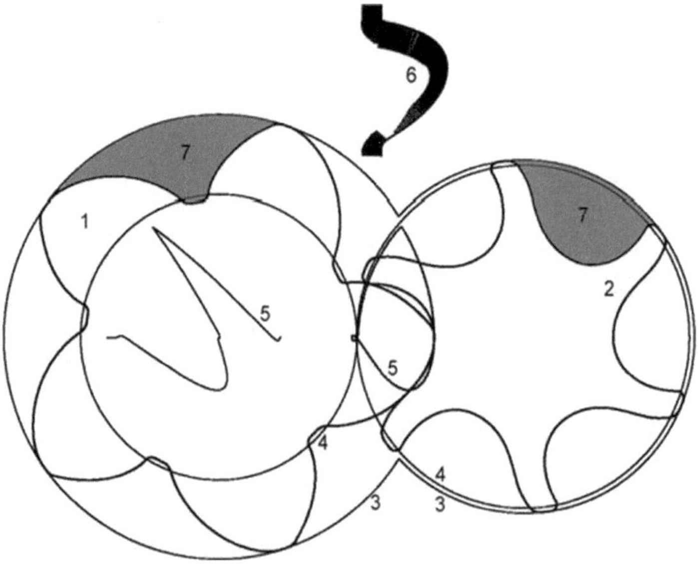

Fig. 1.2. Screw rotor profile: (1) main, (2) gate, (3) rotor external and (4) pitch circles, (5) sealing line, (6) clearance distribution and (7) rotor flow area between the rotors and housing

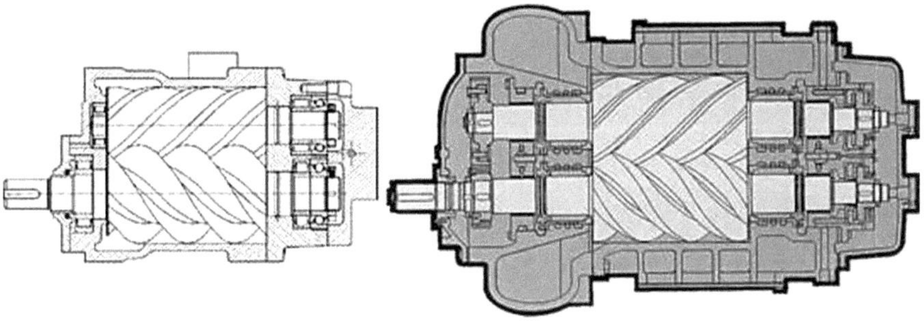

Fig. 1.3. Oil Injected and Oil Free Compressors

Screw machines have a number of advantages over other positive displacement types. Firstly, unlike reciprocating machines, the moving parts all rotate and hence can run at much higher speeds. Secondly, unlike vane machines, the contact forces within them are low, which makes them very reliable. Thirdly, and far less well appreciated, unlike the reciprocating, scroll and vane machines, all the sealing lines of contact which define the boundaries of each cell chamber, decrease in length as the size of the working chamber decreases and the pressure within it rises. This minimises the escape of gas from the chamber due to leakage during the compression or expansion process.

1.2 Types of Screw Compressors

Screw compressors may be broadly classified into two types. These are shown in Fig. 1.3 where machines with the same size rotors are compared:

1.2.1 The Oil Injected Machine

This relies on relatively large masses of oil injected with the compressed gas in order to lubricate the rotor motion, seal the gaps and reduce the temperature rise during compression. It requires no internal seals, is simple in mechanical design, cheap to manufacture and highly efficient. Consequently it is widely used as a compressor in both the compressed air and refrigeration industries.

1.2.2 The Oil Free Machine

Here, there is no mixing of the working fluid with oil and contact between the rotors is prevented by timing gears which mesh outside the working chamber and are lubricated externally. In addition, to prevent lubricant entering the working chamber, internal seals are required on each shaft between the working chamber and the bearings. In the case of process gas compressors, double mechanical seals are used. Even with elaborate and costly systems such as these, successful internal sealing is still regarded as a problem by established process gas compressor manufacturers. It follows that such machines are considerably more expensive to manufacture than those that are oil injected.

Both types require an external heat exchanger to cool the lubricating oil before it is readmitted to the compressor. The oil free machine requires an oil tank, filters and a pump to return the oil to the bearings and timing gear.

The oil injected machine requires a separator to remove the oil from the high pressure discharged gas but relies on the pressure difference between suction and discharge to return the separated oil to the compressor. These additional components increase the total cost of both types of machine but the add on cost is greater for the oil free compressor.

1.3 Screw Machine Design

Serious efforts to develop screw machines began in the nineteen thirties, when turbomachines were relatively inefficient. At that time, Alf Lysholm, a talented Swedish engineer, required a high speed compressor, which could be coupled directly to a turbine to form a compact prime mover, in which the motion of all moving parts was purely rotational. The screw compressor appeared to him to be the most promising device for this purpose and all modern developments in these machines stem from his pioneering work. Typical screw compressor designs are presented in Figs. 1.4 and 1.5. From then until the mid nineteen sixties, the main drawback to their widespread use was the inability to manufacture rotors accurately at an acceptable cost. Two developments then accelerated their adoption. The first was the development of milling machines for thread cutting. Their use for rotor manufacture enabled these components to be made far more accurately at an acceptable cost. The second occurred in nineteen seventy three, when SRM, in Sweden, introduced the "A" profile, which reduced the internal leakage path area, known as the blow hole, by 90%. Screw compressors could then be built with efficiencies approximately equal to those of reciprocating machines and, in their oil flooded form, could operate efficiently with stage pressure ratios of up to 8:1. This was unattainable with reciprocating machines. The use of screw compressors, especially of the oil flooded type, then proliferated.

To perform effectively, screw compressor rotors must meet the meshing requirements of gears while maintaining a seal along their length to minimise leakage at any position on the band of rotor contact. It follows that the compressor efficiency depends on both the rotor profile and the clearances between the rotors and between the rotors and the compressor housing.

Screw compressor rotors are usually manufactured on specialised machines by the use of formed milling or grinding tools. Machining accuracy achievable today is high and tolerances in rotor manufacture are of the order of 5 μm around the rotor lobes. Holmes, 1999 reported that even higher accuracy was

Fig. 1.4. Screw compressor mechanical parts

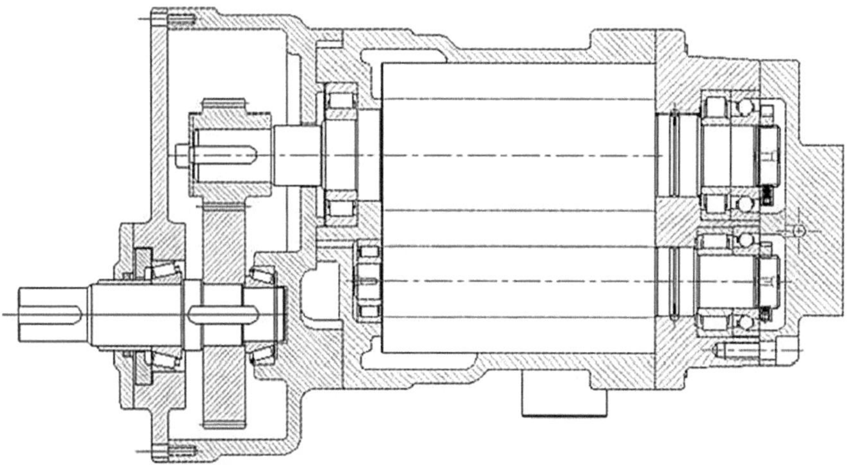

Fig. 1.5. Cross section of a screw compressor with gear box

achieved on the new Holroyd vitrifying thread-grinding machine, thus keeping the manufacturing tolerances within 3 μm even in large batch production. This means that, as far as rotor production alone is concerned, clearances between the rotors can be as small as 12 μm.

Screw machines are used today for different applications both as compressors and expanders. For optimum performance from them a specific design and operating mode is needed for each application. Hence, it is not possible to produce efficient machines by the specification of a universal rotor configuration or set of working parameters, even for a restricted class of machines.

Industrial compressors are required to compress air, refrigerants and process gases. For each application their design must differ to obtain the most desirable result. Typically, refrigeration and process gas compressors, which operate for long periods, must have a high efficiency. In the case of air compressors, especially for mobile applications, efficiency may be less important than size and cost.

Oil free compressed air is delivered almost exclusively by screw compressors. The situation is becoming similar for the case of process gas compression. In the field of refrigeration, reciprocating and vane compressors are continuously being replaced by screw and a dramatic increase in the needs for refrigeration compressors is expected in the next few years.

The range of screw compressors sizes currently manufactured is covered by male rotor diameters of 75 to 620 mm and this permits the delivery of compressed gas flow rates of $0.6\,\mathrm{m}^3/\mathrm{min}$ to $600\,\mathrm{m}^3/\mathrm{min}$. A pressure ratio of 3.5 is attainable in them from a single stage for dry compressors and up to 15 for oil flooded ones. Normal pressure differences are up to 15 bars, but maximum pressure differences sometimes exceed 40 bars. Typically, for oil flooded air compression applications, the volumetric efficiency of these machines now exceeds 90% and the specific power input has been reduced to values which were regarded as unattainable only a few years ago.

1.4 Screw Compressor Practice

The Swedish company SRM was a pioneer and they are still leaders in the field of screw compressor practice. Other companies, like Compair U.K., Atlas-Copco in Belgium, Ingersol-Rand and Gardner Denver in the USA and GHH in Germany follow them closely. York, Trane and Carrier lead in screw compressor applications for refrigeration and air conditioning. Japanese screw compressor manufacturers, like Hitachi, Mycom and Kobe-Steel are also well known. Many relatively new screw compressor companies have been founded in the Middle and Far East. New markets in China and India and in other developing countries open new screw compressor factories. Although not directly involved in compressor production the British company, Holroyd, are the largest screw rotor manufacturer, they are world leaders in tool design and tool machine production for screw compressor rotors.

Despite the increasing popularity of screw compressors, public knowledge and understanding of them is still limited. Three screw compressor textbooks were published in Russian in the early nineteen sixties. Sakun, 1960 gives a full description of circular, elliptic and cycloidal profile generation and a reproducible presentation of a Russian asymmetric profile named SKBK. The profile generation in his book was based on an envelope approach. Andreev,

1961 repeats the theory of screw profiles and makes a contribution to rotor tool profile generation theory. Golovintsov's textbook, 1964, is more general but its section on screw compressors is both interesting and informative. Asomov, 1977, also in Russian, gave a reproducible presentation of the SRM asymmetric profile, five years after it was patented, together with the classic Lysholm profile.

Two textbooks have been published in German. Rinder, 1979, presented a profile generation method based on gear theory to reconstruct the SRM asymmetric profile, seven years after it was patented. Konka, 1988, published some engineering aspects of screw compressors.

Only recently a number of textbooks have been published in English, which deal with screw compressors. O'Neill, 1993, on industrial compressors and Arbon, 1994, on rotary twin shaft compressors. There are a few compressor manufacturers' handbooks on screw compressors and a number of brochures giving useful information on them, but these are either classified or not in the public domain. Some of them, like the SRM Data Book, although available only to SRM licensees, are cited in literature on screw compressors.

There is an extraordinarily large number of patents on screw compressors. Literally thousands have appeared in the past thirty years, of which SRM, alone, holds 750. The patents deal with various aspects of these machines, but especially with their rotor profiles. The SRM patents of Nilson, 1952, for the symmetric profile, Shibbie, 1979, for the asymmetric profile and Astberg 1982, for the "D" profile are the most widely quoted in reference literature on this topic. Ohman, 1999, introduced the "G" profile for SRM. Other examples of successful profiling patents may also be mentioned, namely: Atlas-Copco, Compair with Hough, 1984, Gardner Denver with Edstroem, 1974, Hitachi with Kasuya, 1983, and Ingersoll-Rand with Bowman, 1983. More recently, several highly successful patents were granted to relatively small companies such as Fu Sheng, Lee, 1988, and Hanbel, Chia-Hsing, 1995. A new approach to profile generation, using a rack as the basis for the primary curves, was proposed by Rinder, 1987, and Stosic, 1996.

All patented profiles were generated by a procedure but information on the methods used is hardly disclosed either in the patents or in accompanying publications. Thus it took many years before these procedures became known. Examples of this are: Margolis, who published his derivation of the symmetric circular profile in 1977, 32 years after it had been patented and Rinder, who used gear meshing criteria to reproduce the SRM asymmetric profile in 1979, 9 years after patent publication. It may also be mentioned that Tang, 1995, derived the SRM "D" profile analytically as part of a PhD thesis 13 years after the patent publication. Many other aspects of screw compressors were also patented. These include nearly all their most well known characteristics, such as, oil flooding, the suction and discharge ports following the rotor tip helices, axial force compensation, unloading, the slide valve and the economiser port, most of which were filed by SRM. However, other companies were also keen to file patents. The general impression gained is that patent experts are as important for screw compressor development as engineers.

There is a surprising lack of screw compressor publications in the technical literature. Lysholm's papers in 1942 and 1966 were a mid twentieth century exception, but he did not include any details of the profiling details which he introduced to reduce the blow-hole area. Thus, journal papers like those of Stosic et al., 1997, 1998, may be regarded as an exception. In recent years, publication of screw compressor materials in journals has become more common through the International Institution of Refrigeration Stosic, 1992 and Fujiwara, 1995, the IMechE, with papers by Smith, 1996, Fleming, 1994, 1998 and Stosic 1998, and the ASME, by Hanjalic, 1997. Together these made more information available than the total published in all previous years. Stosic's, 1998 paper is a typical example of the modern practice of timely publishing.

There are three compressor conferences which deal exclusively or partly with screw compressors. These are the biennial compressor technology conference, held at Purdue University in the USA, the IMechE international conference on compressors and their systems, in England and the "VDI Schraubenkompressoren Tagung" in Dortmund, Germany. Despite the number of papers on screw compressors published at these events, only a few of them contain useful information on rotor profiling and compressor design. Typical Purdue papers cited as publications from which a reader can gain information on this are: Edstroem, 1992, Stosic, 1994 and Singh, 1984, 1990. Zhang, 1992, indicates that they used envelope theory to calculate some geometric features of their rotors. The Dortmund proceedings give some interesting papers such as that by Rinder, 1984, who presented the rack generation of a screw rotor profile, including a fully reproducible pattern based on gear theory. Hanjalic, 1994 and Holmes, 1994, give more details on profiling, manufacturing and control. Kauder, 1994, 1998 and Stosic, 1998 are typical examples of successful university reseach applied to real engineering. Sauls, 1994, 1998, may be regarded as an example of fine engineering work. The London compressor conference included some interesting papers like those of Edstroem, 1989 and Stosic et al., 1999.

Many reference textbooks on gears give useful background for screw rotor profiling. However all of them are limited to the classical gear conjugate action condition. Litvin, 1968 and 1956–1994 may be regarded as an exception to this practice, in giving gearing theory which can be applied directly to screw compressor profiling.

1.5 Recent Developments

The efficient operation of screw compressors is mainly dependent on proper rotor design. An additional and important requirement for the successful design of all types of compressor is an ability to predict accurately the effects on performance of the change in any design parameter such as clearance, rotor profile shape, oil or fluid injection position and rate, rotor diameter and proportions and speed.

Now, when clearances are tight and internal leakage rates have become small, further improvements are only possible by the introduction of more refined design principles. The main requirement is to improve the rotor profiles so that the internal flow area through the compressor is maximised while the leakage path is minimised and internal friction, due to relative motion between the contacting rotor surfaces, is made as small as possible.

Although it may seem that rotor profiling is now in a fully developed state, this is far from true. In fact there is room for substantial improvement. The most promising seems to be through rack profile generation which gives stronger but lighter rotors with higher throughput and lower contact stress. The latter enables lower viscosity lubricant to be used.

Rotor housings with better shaped ports can be designed using a multivariable optimization technique. This reduces flow losses thus permitting higher rotor speeds and more compact machines.

Improvements in compressor bearing design achieved in recent years now enable process fluid lubrication in some cases. Also seals are more efficient today. All these give scope for more effective and more efficient screw compressors.

1.5.1 Rotor Profiles

The practice predominantly used for the generation of screw compressor rotor profiles is to create primary profile curves on one of the real screw rotors and to generate a corresponding secondary profile curve on the other rotor using some appropriate conjugate action criterion. Any curve can be used as a primary one, but traditionally the circle is the most commonly used. All circles with centres on the pitch circles generate a similar circle on another rotor. It is the same if the circle centres are at the rotor axes.

Circles with centres offset from the pitch circles and other curves, like ellipses, parabolae and hyperbolae have elaborate counterparts. They produce generated curves, so called trochoids, on the other rotor. Similarly, points located on one rotor will cut epi- or hypocycloids on the other rotor. For decades the skill needed to produce rotors was limited to the choice of a primary arc which would enable the derivation of a suitable secondary profile.

The symmetric circular profile consists of circles only, Lysholm's asymmetric profile, apart from pitch circle centered circles, introduced a set of cycloids on the high pressure side, forming the first asymmetric screw rotor profile. The SRM asymmetric profile employs an offset circle on the low pressure side of the gate rotor, followed later by the SKBK profile introducing the same on the main rotor. In the both cases the evolved curves were given analytically as epi- or hypocycloids. The SRM "D" profile consists exclusively of circles, almost all of them eccentrically positioned on the main or gate rotor. All patents following, give primary curves on one rotor and secondary, generated curves on the other rotor, all probably based on derivations of classical gearing or some other similar condition. More recently, the circles have been gradually replaced by other curves, such as elipses in the FuSheng profiles, parabolae in the Compair and Hitachi profiles and hyperbolae in the "hyper" profile. The hyperbola in the latest profiles seems to be the most appropriate replacement giving the best ratio of rotor displacement to sealing line length.

Another practice to generate screw rotor profile curves is to use imaginary, or "non-physical" rotors. Since all gearing equations are independent of the coordinate system in which they are expressed, it is possible to define primary arcs as given curves using a coordinate system which is independent of both rotors. By this means, in many cases the defining equations may be simplified. Also, the use of one coordinate system to define all the curves further simplifies the design process. Typically, the template is specified in a rotor independent coordinate system. The same is valid for a rotor of infinite radius which is a rack. From this, a secondary arc on some of the rotors is obtained by a procedure, which is called "rack generation". The first ever published patent on rack generation by Menssen, 1977, lacks practicality but conveniently uses the theory. Rinder, 1987 and recently Stosic, 1996 give a better basis for profile generation.

An efficient screw compressor needs a rotor profile which has a large flow cross section area, a short sealing line and a small blow-hole area. The larger the cross section area the higher the flow rate for the same rotor sizes and rotor speeds. Shorter sealing lines and a smaller blow-hole reduce leakages. Higher flow and smaller leakage rates both increase the compressor volumetric efficiency, which is the rate of flow delivered as a fraction of the sum of the flow plus leakages. This in turn increases the adiabatic efficiency because less power is wasted in the compression of gas which is recirculated internally.

The optimum choice between blow hole and flow areas depends on the compressor duty since for low pressure differences the leakage rate will be relatively small and hence the gains achieved by a large cross section area may outweigh the losses associated with a larger blow-hole. Similar considerations determine the best choice for the number of lobes since fewer lobes imply greater flow area but increased pressure difference between them.

As precise manufacture permits rotor clearances to be reduced, despite oil flooding, the likelihood of direct rotor contact is increased. Hard rotor contact leads to deformation of the gate rotor, increased contact forces and ultimately rotor seizure. Hence the profile should be designed so that the risk of seizure is minimised.

The search for new profiles has been both stimulated and facilitated by recent advances in mathematical modelling and computer simulation. These analytical methods may be combined to form a powerful tool for process analysis and optimisation and thereby eliminate the earlier approach of intuitive changes, verified by tedious trial and error testing. As a result, this approach to the optimum design of screw rotors lobe profiles has substantially evolved over the past few years and is likely to lead to further improvements in machine performance in the near future. However, the compressor geometry and the processes involved within it are so complex that numerous approximations are required for successful modelling. Consequently, the computer models and numerical codes reported in the open literature often differ in their approach and in the mathematical level at which various phenomena are modelled. A lack of comparative experimental verification still hinders a comprehensive validation of the various modelling concepts. In spite of this, computer modelling and optimization are steadily gaining in credibility and are increasingly employed for design improvement.

The majority of screw compressors are still manufactured with 4 lobes in the main rotor and 6 lobes in the gate rotor with both rotors of the same outer diameter. This configuration is a compromise which has favourable features for both, dry and oil-flooded compressor applications and is used for air and refrigeration or process gas compressors. However, other configurations, like 5/6 and 5/7 and recently 4/5 and 3/5 are becoming increasingly popular. Five lobes in the main rotor are suitable for higher compressor pressure ratios, especially if combined with larger helix angles. The 4/5 arrangement has emerged as the best combination for oil-flooded applications of moderate pressure ratios. The 3/5 is favoured in dry applications, because it offers a high gear ratio between the gate and main rotors which may be taken advantage of to reduce the required drive shaft speed.

Figure 1.6 shows pairs of screw compressor rotors plotted together for comparison. They are described by their commercial name or by a name which denotes their patent.

The first group gives rotors with 4 lobes on the main and 6 lobes on the gate rotor. This rotor configuration is the most universally acceptable for almost any application. The SRM asymmetric profile Shibbie, 1979, which historically appears to be the most successful screw compressor profile is near the top.

The next one is Astberg's SRM "D" profiles 1982.

The largest group of rotors presented is in 5/6 configuration which is becoming the most popular rotor combination because it combines a large displacement with large discharge ports and favourable load characteristics in a small rotor size. It is equally successful in air compression and in refrigeration and air-conditioning. The group starts with the SRM "D" profile, followed by the "Sigma", Bammert, 1979 profile, the FuSheng, Lee, 1988 and the "Hyper", Chia-Hsing, 1995 profile. All the profiles shown are "rotor-generated" and the main difference between them is in the leading lobe which is an offset circle on the main, a circle followed by a line, an ellipse and a hyperbola respectively. The hyperbola appears to be the best possible geometrical solution for that purpose. The last two are the rack generated rotors of Rinder, 1984 and Stosic, 1996. The primary curves for these were selected and distributed on a rack to create a larger cross section area with stronger gate rotor lobes than in any other known screw compressor rotor.

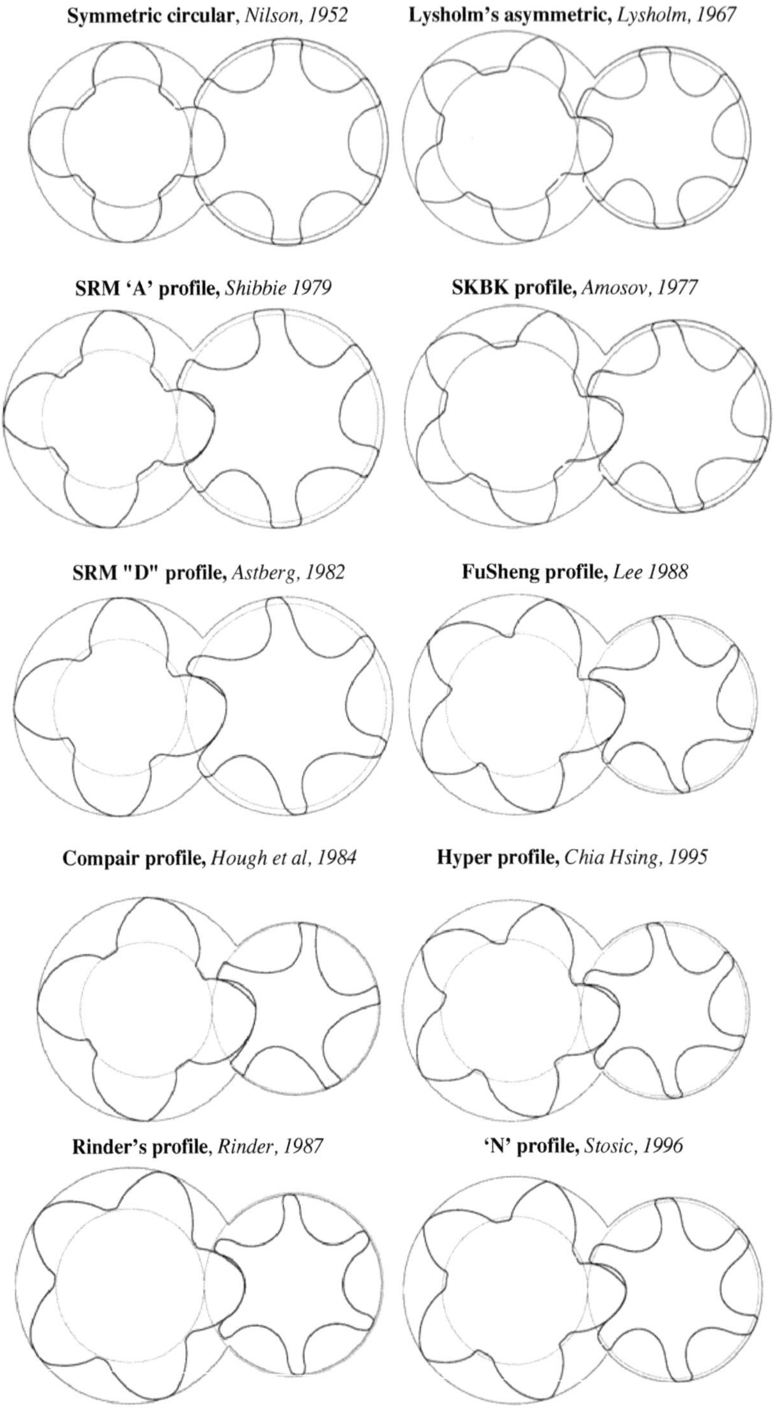

Fig. 1.6. Popular screw compressor rotors

Two additional favourable features characterize the “N” profile rotors. They maintain a seal over the entire contact length while maintaining a small blow-hole. This was not the case with Rinder, 1984. Additionally, the two contact bands in the proximity of the pitch circles are straight lines on the rack. These form involutes on the rotors. Hence the relative motion between the rotors is the best possible.

1.5.2 Compressor Design

The SRM Asymmetric profile, which was responsible for the biggest single advance in screw compressor performance, was achieved by replacing the high pressure face of the circular lobes by a system of cycloids. This led to a substantial reduction in the blow-hole area, and a corresponding increase in delivery rate and efficiency. In addition, the low pressure face of the lobes was modified through the eccentricity of the gate rotor circle, which resulted in a different shape for the corresponding lobe face of the main rotor. The resulting SRM “A” profile replaced all others and is still widely used for a number of applications.

Although advanced rotor profiles are needed for further improvements in screw compressor efficiency, all other components must then be designed to take advantage of their potential if the full performance gains are to be achieved. Thus rotor to housing clearances, especially at the high pressure end must be properly selected. This in turn requires either expensive bearings with smaller clearances or cheaper bearings with their clearances reduced to an acceptable value by preloading.

A screw compressor, especially of the oil flooded type, which operates with high pressure differences, is heavily loaded by axial and radial forces, which are transferred to the housing by the bearings. Rolling element bearings are normally chosen for small and medium screw compressors and these must be carefully selected to obtain a satisfactory design. Usually two bearings are employed on the discharge end of the rotor shafts in order to absorb the radial and axial loads separately. Also the distance between the rotor centre lines is in part determined by the bearing size and internal clearance.

The contact force between the rotors, which is determined by the torque transferred between the rotors, plays a key role in compressors with direct rotor contact. The contact force is relatively small in the case of a main rotor driven compressor. In the case of a gate rotor drive, the contact force is substantially larger and this case should be excluded from any serious consideration.

The same oil is used for oil flooding and for bearing lubrication but the supply to and evacuation from the bearings is separate to minimise the friction losses. Oil is injected into the compressor chamber at the place where thermodynamic calculations show the gas and oil inlet temperature to coincide. The position is defined on the rotor helix with the injection hole located so that the oil enters tangentially in line with the gate rotor tip in order to recover as much as possible of the oil kinetic energy.

To minimise the flow losses in the suction and discharge ports, the following features are included. The suction port is positioned in the housing to let the gas enter with the fewest possible bends and the gas approach velocity is kept low by making the flow area as large as possible. The discharge port size is first determined by estimating the built-in-volume ratio required for optimum thermodynamic performance. It is then increased in order to reduce the gas exit velocity and hence obtain the minimum combination of internal and discharge flow losses.

The casing should be carefully dimensioned to minimize its weight, containing reinforcing bars across the suction port to improve its rigidity at higher pressures.

The screw compressor is a very simple machine with only a few moving parts, comprising two rotors and usually four to six bearings. Its design has evolved into very typical forms.

The tendency is to design as small as possible a machine to produce and perform satisfactorily. This means that the rotor tip speed is limited by the need to obtain an acceptable machine efficiency. It is almost a rule to use rolling element bearings wherever possible to take advantage of their smaller clearances compared with journal bearings. The ports are made as wide as possible to minimize suction and discharge gas speeds. All this results in screw compressor designs with little variation.

It must be added that recent advances in the development of low friction rolling element bearings have greatly contributed to the improvements in screw compressor efficiency.

Such guidelines are also essential for further improvement of existing screw compressor designs, and broadening the range of uses for these machines.

2

Screw Compressor Geometry

To be able to predict the performance of any type of positive displacement compressor it is necessary to have a facility to estimate the working chamber size and shape at any point in its operating cycle.

In the case of screw compressors this implies the need to be able to define the rotor lobe profiles, together with any additional parameters needed for the rotor and housing geometry to be fully specified. A set of subprograms which can compute the lobe profiles and the complete geometry of the working space of a screw machine of almost arbitrary design has been developed. The default version is a new asymmetric profile, called Demonstrator, which can model any realistic combination of numbers of lobes in the main and gate screw rotors. However, any other known or even a completely new profile can be generated, with little or no modification of the code. Such profiles must, of course, satisfy geometrical constraints in order to obtain a realistic solution.

2.1 The Envelope Method as a Basis for the Profiling of Screw Compressor Rotors

The envelope method is used here as a basis for the generation of a screw compressor rotor profile. The method states that two surfaces are in mesh if each generates or envelops the other under a specified relative motion. This approach is becoming increasingly popular, and details of how it may be applied to screw compressor rotor profiles, have been given by Stosic 1998. Although the generation of screw compressor rotors can be regarded as a two-dimensional problem, a three-dimensional approach is given here as a general starting point. This also gives an opportunity to generate the rotor tools from the same equations. Andreev, 1961 and, more recently, Xing, 2000a, also use the envelope method for screw compressor rotor generation. A similar approach was applied by Tang, 1995. The derivation of the gearing procedure based on the envelope method is presented in full in Appendix A. The envelope method is also used as a basis for the generation of the rotor manufacturing tools.

2.2 Screw Compressor Rotor Profiles

Screw machine rotors have parallel axes and a uniform lead and they are therefore a form of helical gears. As shown in Fig. 2.1, the rotor centre distance is $C = r_{1w} + r_{2w}$, where r_{1w} and r_{2w} are the radii of the main and gate rotor pitch circles respectively. The rotors make line contact and the meshing criterion in the transverse plane perpendicular to their axes is the same as that of spur gears. Although spur gear meshing fully defines helical screw rotors, it is more convenient to use the envelope condition for crossed helical gears and simplify it by setting the shaft angle between the two rotor axes, $\Sigma = 0$, to get the required meshing condition.

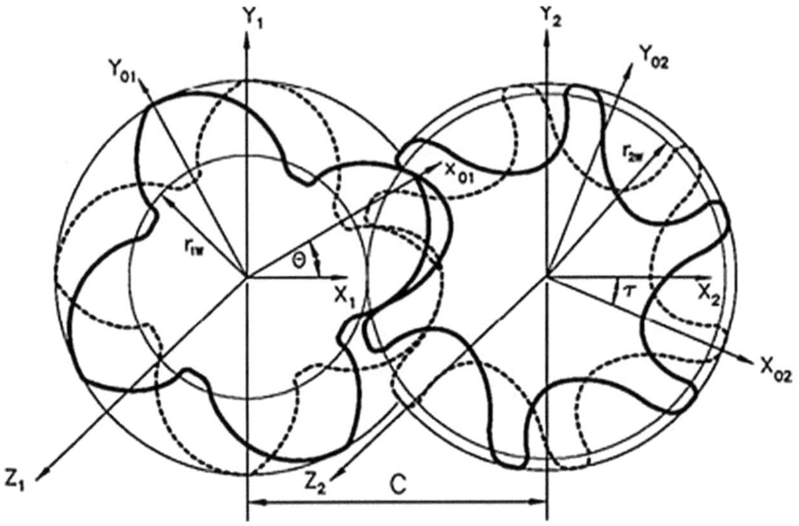

Fig. 2.1. Screw compressor rotors with parallel shafts and their coordinate systems

To start the procedure of rotor profiling, the profile point coordinates in the transverse plane of one rotor, x_{01} and y_{01} and their first derivatives, either $\frac{\partial x_{01}}{\partial t}$ and $\frac{\partial y_{01}}{\partial t}$ or $\frac{dy_{01}}{dx_{01}}$ must be known. This profile may be specified on either the main or gate rotors or in sequence on both. Also the primary profile may be defined as a rack.

Since $\Sigma = 0$, the general meshing condition presented in Appendix A reduces for the screw machine rotor to:

$$\frac{dy_{01}}{dx_{01}}\left(ky_{01} - \frac{C}{i}\sin\theta\right) + kx_{01} + \frac{C}{i}\cos\theta = 0 \tag{2.1}$$

where $i = p_2/p_1$ and $k = 1-1/i$. This equation can be solved only numerically.

θ still can be obtained only numerically. Once obtained, the distribution of θ along the profile may be used to calculate the meshing rotor profile point coordinate, as well as to determine the sealing lines and paths of proximity between the two rotors. The rotor rack coordinates may also be calculated from the same θ distribution.

Since $\tau = \theta/i$ for parallel axes, the meshing profile equations of the gate rotor in the transverse plane are calculated as:

$$\begin{aligned} x_{02} &= x_{01}\cos k\theta - y_{01}\sin k\theta - C\cos\frac{\theta}{i} \\ y_{02} &= x_{01}\sin k\theta + y_{01}\cos k\theta + C\sin\frac{\theta}{i} \end{aligned} \tag{2.2}$$

The rack coordinates can be obtained uniquely if the rack-to-rotor gear ratio i tends to infinity:

$$\begin{aligned} x_{0r} &= x_{01}\cos\theta - y_{01}\sin\theta \\ y_{0r} &= x_{01}\sin\theta + y_{01}\cos\theta - r_1\theta \end{aligned} \tag{2.3}$$

Conversely, if the gate rotor curves are given, the generated curves will be placed on the gate rotor and similar equations with substituted indices will be used to generate the main rotor profile.

However, if the primary curves are given on the rack, their coordinates x_{0r} and y_{0r}, as well as their first derivatives, $\frac{\partial x_{0r}}{\partial t}$ and $\frac{\partial y_{0r}}{\partial t}$, or $\frac{dy_{0r}}{dx_{0r}}$ should be known and the generated curves will be calculated at the rotors as:

$$\begin{aligned} x_{01} &= x_{0r}\cos\theta - (y_{0r} - r_{1w})\sin\theta \\ y_{01} &= x_{0r}\sin\theta + (y_{0r} - r_{1w})\cos\theta \end{aligned} \tag{2.4}$$

after the meshing condition is obtained from:

$$\frac{dy_{0r}}{dx_{0r}}(r_{1w}\theta - y_{0r}) - (r_{1w} - x_{0r}) = 0 \tag{2.5}$$

The rack meshing condition θ can be solved directly and does not require a numerical procedure for its evaluation, this is another advantage of the rack generation procedure.

The procedure given to solve the meshing condition either numerically or directly permits the introduction of a variety of primary arc curves and effectively constitutes a general procedure. Further on, the numerical derivation of the primary arcs enables a fully general approach when only the coordinates of the primary curves need be known without their derivatives. In such a case, any analytical function and even discrete point functions can be used as primary arcs. The approach adopted further simplifies the procedure. Only the primary arcs need be given, since the secondary ones are not derived but are evaluated automatically by means of the numerical procedure.

The sealing line of screw compressor rotors is somewhat similar to the gear contact line. Since there exists a clearance gap between rotors, the sealing line is a line consisting of points of the most proximate rotor position. Its coordinates are x_1, y_1 and z_1 and they are calculated for the same θ distribution.

The most convenient practice to obtain an interlobe clearance gap is to consider the gap as the shortest distance between two rotor racks of the main and gate rotor sealing points in the cross section normal to the rotor helicoids.

The rotor racks, obtained from the rotors by the reverse procedure, can include all manufacturing and positioning imperfections. Therefore the resulting clearance distribution may represent real life compressor clearances. From normal clearances, a transverse clearance gap may be obtained by the appropriate transformation.

Further rearrangement of the rotor meshing condition gives a form which is frequently used for the profiling of spur and helical gears. Let ϕ be a pressure angle, or angle of the normals of the rotors and the rack at the contact point. For the given rotor coordinates x_{01} and y_{01}, $\tan\theta_1 = -\frac{dy_{01}}{dx_{01}}$. Let φ be a profile angle at the contact point given as $\tan\varphi_1 = \frac{y_{01}}{x_{01}}$. This implies that $x_{01} = r_{01}\cos\varphi_1$ and $y_{01} = r_{01}\sin\varphi_1$, where r_{01} is a point radius. Let θ be the meshing condition, or the rotation angle of the main rotor for which the rotors and rack are in contact at local points (x_{01}, y_{01}), (x_{02}, y_{02}) and (x_{0r}, y_{0r}). The relation between these three angles required for calculation of the meshing condition θ is then obtained as:

$$\frac{\sin(\theta_1 + \theta)}{r_{01}} = \frac{\sin(\theta_1 - \varphi_1)}{r_{1w}} \tag{2.6}$$

A graphical presentation of this meshing condition given in Fig. 2.2 confirms that the Willis gearing condition, which states that the normals of both, the gears and the rack at their contact point pass through the pitch point, is only a special case of the envelope meshing condition.

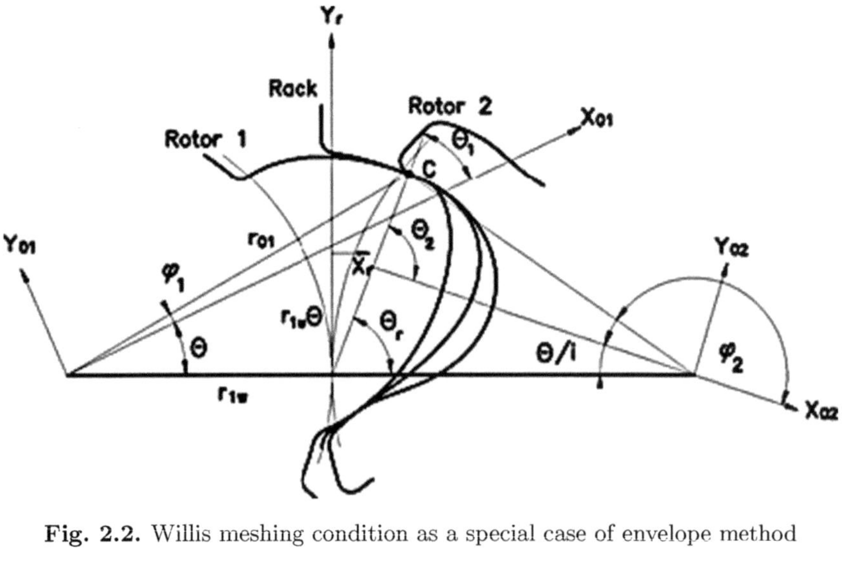

Fig. 2.2. Willis meshing condition as a special case of envelope method

2.3 Rotor Profile Calculation

For a further analysis of the compressor geometry, several generic definitions are introduced here. The rotor gear ratio is $i = \frac{r_{2w}}{r_{1w}} = \frac{z_2}{z_1}$, where z_1 and z_2 are the numbers of lobes on the main and gate rotor. Since the screw compressor rotors are three-dimensional bodies, a helix angle ψ is defined at the rotor radius, while ψ_w corresponds to the pitch circle, $\frac{tg\psi}{tg\psi_w} = \frac{r}{r_w}$. The helix angle defines the rotor lead h, which can be given relative to the unit angle $p = \frac{h}{2\pi}$. The rotor length L, the wrap angle φ and the lead are interrelated $\frac{L}{\varphi} = \frac{h}{2\pi} = p$. If the rotors are unwrapped, a simple relation between the wrap and helix angles can be established, $tg\psi_w = \frac{\varphi r_w}{h} = \frac{2\pi r_w}{L}$. The lead angle is the complement of the helix angle.

As shown in Fig. 2.3, the rotor displacement is the product of the rotor length and its cross section area, which is denoted by the number 1, while the overlapping areas on the main and gate rotors are denoted by the number 2.

2.4 Review of Most Popular Rotor Profiles

This section reviews a procedure to calculate various screw profiles. Initially a detailed presentation of rotor creation by the rotor generation procedure is given. The rotor profile in this case is a very simple hypothetical one. It has been applied in practice, but also been frequently used for training purposes. Furthermore, this profile may be very conveniently used as a basis for individual development of screw compressor rotors and such use is encouraged here. Based on this, other profiles are briefly derived, like the early SKBK profile, the "Sigma" profile by Kaeser, the "Hyper" profile by Hanbel and the Fu Sheng and Hitachi profiles. Also the symmetric profile and asymmetric

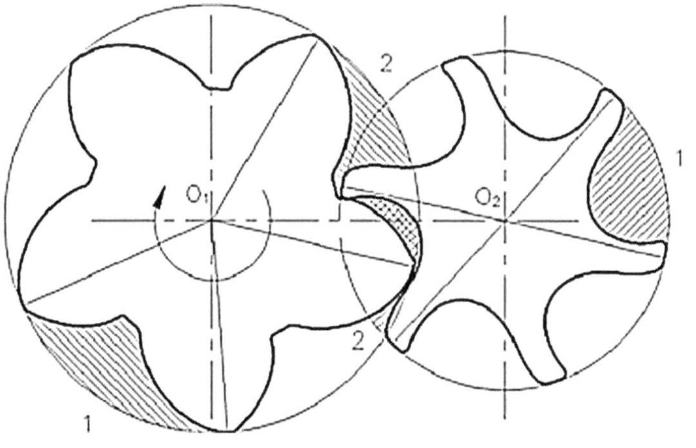

Fig. 2.3. Rotor cross section area and overlapping sectors

"A", "D" and "G" profiles of SRM and the "Cyclon" profile by Compair are reviewed. Finally, two rack generated profiles are described namely the "N" and Rinder's profile.

2.4.1 Demonstrator Rotor Profile ("N" Rotor Generated)

The demonstrator profile is a rotor generated "N" profile and is not to be confused with the patented rack generated "N" profile. The primary or generating lobe profile of the Demonstrator is given on the main rotor and the profile is divided into several segments. The division between the profile segments is denoted by capital letters and each segment is defined separately by its characteristic angles, as shown in the Fig. 2.4. The lobe segments of this profile are essentially parts of circles on one rotor and curves corresponding to the circles on the opposite rotor.

A graphical presentation of this profile is presented in Fig. 2.5. The following summarizes the specific expressions for the x-y coordinates of the lobe profiles of the main screw rotor, with respect to the centre of the rotor O_1. Given are the pitch radii, r_{1w} and r_{2w} and the rotor radii r, r_0, r_2, r_3 and r_4. The external and internal radii are calculated as $r_{1e} = r_{1w} + r$ and $r_{1i} = r_{1w} - r_0$, as well as $r_{1e} = r_{1w} + r$ and $r_{1i} = r_{1w} - r_0$ for the main and gate rotor respectively.

In the demonstrator profile, segment A_1B_1 is a circle of radius r_1 on the main rotor. The angular parameter t varies between $-\theta_1 < t < 0$.

$$x_{01} = (r_{1e} - r_1) + r_1 \cos t$$
$$y_{01} = r_1 \sin t \tag{2.7}$$

Fig. 2.4. Demonstrator Profile

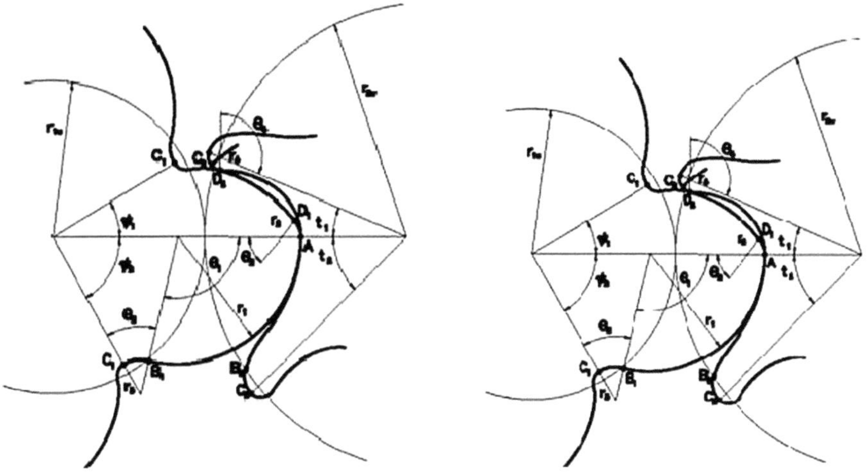

Fig. 2.5. Details of the Demonstrator Profile

r_{1e} is given, while r_1 and θ_1 are calculated through the following procedure, which is presented graphically in Fig. 2.5. There the flat side of the profile is presented in the position where points F_1 and F_2 coincide:

$$\cos\theta_r = -\frac{r_{1w}^2 + r_{C2}^2 - (r_{1e} - r_2)^2}{2r_{1w}r_{C2}} = \frac{r_{2w}^2 + r_{C2}^2 - (r_{2e} - r_4)^2}{2r_{2w}r_{C2}} \tag{2.8}$$

After θ and r_{C2} are obtained from these equations, θ_1 and θ_4 can be calculated as:

$$\begin{aligned}\cos\theta_1 &= \frac{(r_{1e} - r_2)^2 + r_{C2}^2 - r_{1w}^2}{2(r_{1e} - r_2)\,r_{C2}} \\ \cos\theta_4 &= -\frac{(r_{2e} - r_4)^2 + (r_{C2} + r_4 + r_2)^2 - r_{2w}^2}{2(r_{2e} - r_4)(r_{C2} + r_4 + r_2)}\end{aligned} \tag{2.9}$$

The other angles are: $\theta = \theta_r - \theta_1$ and $t_1 + \theta/i = \theta_4 - \theta_r$.

On the round side of the rotors, $\psi_2 = 2\pi/z_1 - \psi_1$, where z_1 is the number of lobes in the main rotor. The radius r_1 is now calculated from:

$$2(r_{1i} + r_3)(r_{1e} - r_1)\cos\psi_2 = (r_{1i} + r_3)^2 + (r_{1e} - r_1)^2 - (r_1 + r_3)^2 \tag{2.10}$$

Other necessary angles are calculated as follows:

$$\frac{\sin\theta_3}{r_{1e} - r_1} = \frac{\sin\psi_2}{r_1 + r_3} \tag{2.11}$$

The segment B_1C_1 is on a circle of radius r_3 on the main rotor, where $\pi - \psi_1 < t < \pi - \theta_1$.

$$\begin{aligned} x_{01} &= (r_{1i} - r_3)\cos\psi_2 + r_3 \cos t \\ y_{01} &= -(r_{1i} - r_3)\sin\psi_2 + r_3 \sin t \end{aligned} \tag{2.12}$$

Profile portion A_1D_1 is a circle of radius r_2 on the main rotor, $0 < t < \theta_2$.

$$\begin{aligned} x_{01} &= r_{1e} - r_2 \cos t \\ y_{01} &= r_2 \sin t \end{aligned} \tag{2.13}$$

Segment C_1D_1 emerges as a trochoid on the main rotor generated by the circle of radius r_4 on the gate rotor, $-\theta_4 - \tau_1 < t < -\pi - \tau_1$. The trochoid is obtained from the gate rotor coordinates through the same meshing procedure. The circle C_2D_2 is:

$$\begin{aligned} x_{02} &= (r_{2e} - r_4)\cos\tau_1 + r_4 \cos t \\ y_{02} &= (r_{2e} - r_4)\sin\tau_1 + r_4 \sin t \end{aligned} \tag{2.14}$$

Now, when all the segments of the main rotor are known, they are used as source curves. The gate rotor lobe can now be generated completely by the meshing procedure described in the previous section.

Although essentially simple, the Demonstrator profile contains all the features which characterize modern screw rotor profiles. The pressure angles on both, the flat and the round profile lobes are not zero. This is essential for successful manufacturing. The profile is generated by the curves and not by points. This further enhances its manufacturability. By changing its parameters, C, r, r_0, r_2, r_3 and r_4, a variety of profiles can be generated, some with positive gate rotor torque, some suitable for low pressure ratios, and others for high pressure ratio compression. The profile is fully computerized and can be used for demonstration, teaching and development purposes.

2.4.2 SKBK Profile

Amosov's 1977 SKBK profile is the first modern Russian profile to be published in the open literature and it is shown in Fig. 2.6. The profile has the same layout and sequence of segments as the Demonstrator profile apart from the fact that the circles r_2 and r_3 the substituted by cycloids and the segments AB and AF are generated by point generation. This can be readily achieved if r_2 and r_3 in the Demonstrator profile tend to zero.

Similarly to the Demonstrator profile, SKBK profile has an eccentric circle on the round lobe of the main rotor, which gives a pressure angle far different from zero in the pitch circle area. This further ensures both its ease of manufacture and the gate rotor torque stability. This characteristic of the SKBK profile was published at least five years prior the SRM "D" rotor patents which claimed the same feature. However, since the flat lobe sides on the main and gate rotors are generated by points E and A on the gate and main rotor respectively and since E is positioned on the gate rotor pitch circle, the pressure angle at the pitch circle on the flat side is zero. This does not allow manufacturing of this profile by milling or grinding unless the profile is modified.

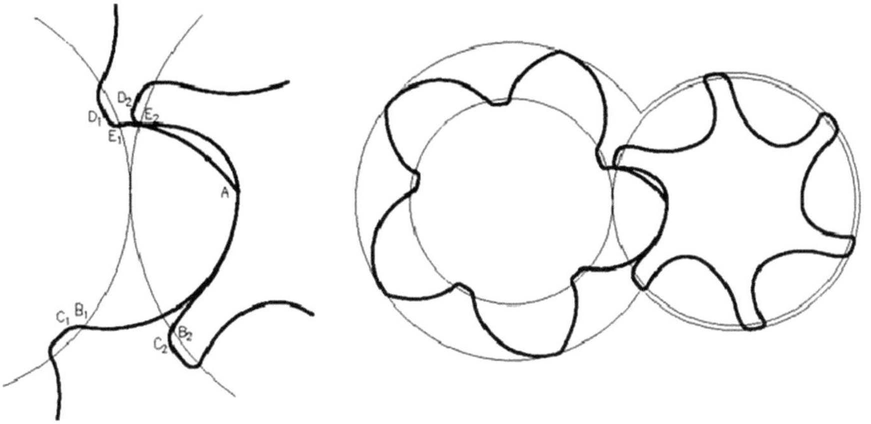

Fig. 2.6. SKBK Profile

Fig. 2.7. Fu Sheng Profile

2.4.3 Fu Sheng Profile

The Fu Sheng profile, as shown in Fig. 2.7, is practically the same as the Demonstrator, but has one distinguishing feature. The segment AB is an ellipse.

2.4.4 "Hyper" Profile

The "Hyper" profile is virtually the same as the Fu Sheng profile, apart from the segment AB, which is a hyperbola on the main rotor instead of the ellipse of the original Fu Sheng profile. However, despite such a small difference, the "Hyper" is a better profile giving larger screw compressor displacement, a shorter sealing line and stronger gate rotor lobes. The Hitachi profile has the same layout as the "Hyper" profile.

2.4.5 "Sigma" Profile

The "Sigma" is a relatively old profile. It was developed in the late nineteen seventies as a response to SRM awarding an exclusive licence to Aerzener in Germany. Other German manufacturers, such as GHH and Kaeser, therefore, needed to develop their own profiles. The "Sigma", shown in Fig. 2.8 is a beautiful and efficient profile. However, new and better profiles are now available. The flat side of the "Sigma" lobe is the same as that of the Demonstrator profile, but the round side of the profile is generated from the flat side by an envelope of circles, which touch both the flat and the round sides, the radii of which are given in advance. This is an acceptable method of profile generation if nothing more general is known, but seriously limits the generation procedure. There are several modifications of the "Sigma" profile. One of these, which is presented here, comprises a straight line BC_2 on the round side of the gate rotor. This modification significantly improves the profile, which is less limited than the original.

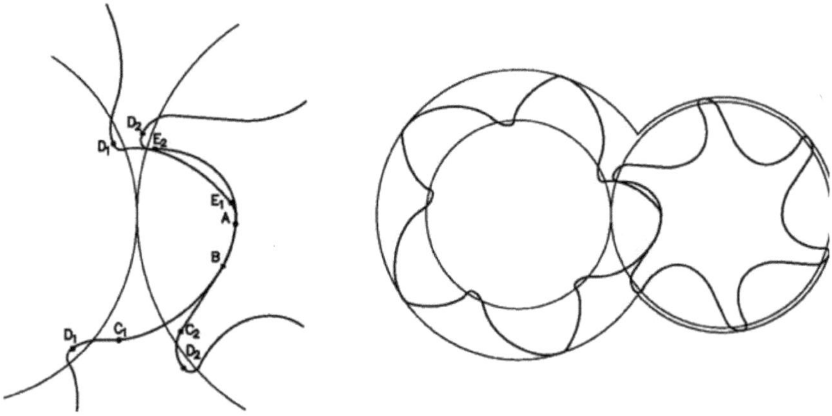

Fig. 2.8. Sigma Profile

2.4.6 "Cyclon" Profile

The "Cyclon" shown in Fig. 2.9 is a profile developed by Compair. The layout and sequence of profile segments are not so different from the Demonstrator, but the "Cyclon" introduces parabolae instead of circles in segments BC, GH and JH. One of the interesting features of the "Cyclon" profile is the "negative" torque on the gate rotor which results in rotor contact on the flat side of the rotors.

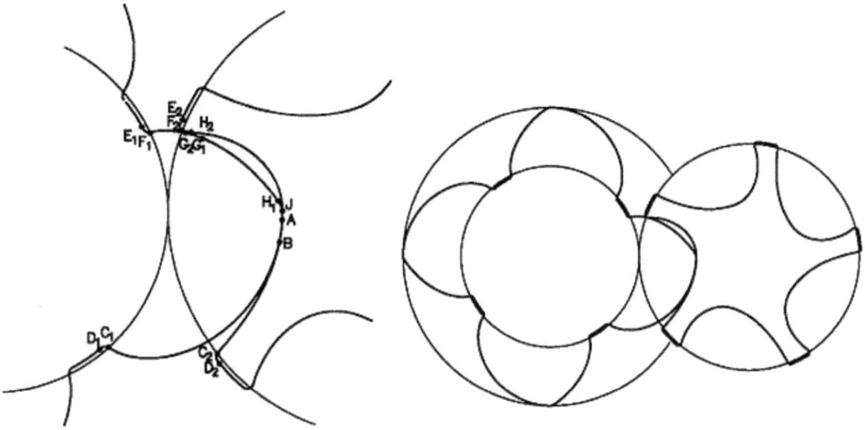

Fig. 2.9. Cyclon Profile

2.4.7 Symmetric Profile

The Symmetric profile, shown in Fig. 2.10 is very simple and consists of three circles on the main rotor with centres positioned either on the rotor centre or on the pitch circle of the main rotor. Since the circles are on the main rotor with centres either at the rotor centre or on the pitch circle, they only generate circles on the gate rotor with centres either in the rotor centre, or on the rotor pitch circle. Is is therefore not surprising that this was the first screw rotor profile ever generated.

Segment D_1E_1 is a circle of radius $r_{1w} - r_0$ with its centre on the rotor axis, while segment E_1F_1 is a circle of radius r_0. Segment F_1A_1 is on a circle of radius r. Both, the last two segments have their centres on the rotor pitch circle. Further segments are symmetrically similar to the given ones.

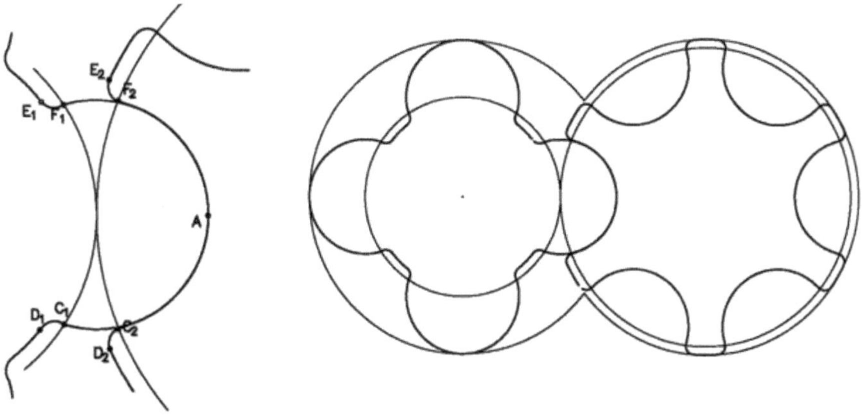

Fig. 2.10. Symmetric Circular Profile

The Symmetric profile has a huge blow-hole area which excludes it from any compressor application where a high or even moderate pressure ratio is involved. However, the symmetric profile performs surprisingly well in low pressure compressor applications.

More details about the circular profile can be found in Margolis, 1978.

2.4.8 SRM "A" Profile

The SRM "A" profile is shown in Fig. 2.11. It retains all the favourable features of the symmetric profile like its simplicity while avoiding its main disadvantage, namely, the large blow-hole area. The main goal of reducing the blow hole area was achieved by allowing the tip points of the main and gate rotors to generate their counterparts, trochoids on the gate and main rotor respectively. The "A" profile consists mainly of circles on the gate rotor and one line which passes through the gate rotor axis.

The set of primary curves consists of: D_2C_2, which is a circle on the gate rotor with the centre on the gate pitch circle, and C_2B_2, which is a circle on the gate rotor, the centre of which lies outside the pitch circle region. This was a new feature which imposed some problems in the generation of its main rotor counterpart, because the mathematics used for profile generation at that time was insufficient for general gearing. This eccentricity ensured that the pressure angles on the rotor pitches differ from zero, resulting in its ease of manufacture. Segment BA is a circle on the gate rotor with its centre on the pitch circle. The flat lobe sides on the main and gate rotors were generated as epi/hypocycloids by points G on the gate and H on the main rotor respectively. GF_2 is a radial line at the gate rotor. This brought the same benefits to manufacturing as the previously mentioned circle eccentricity on

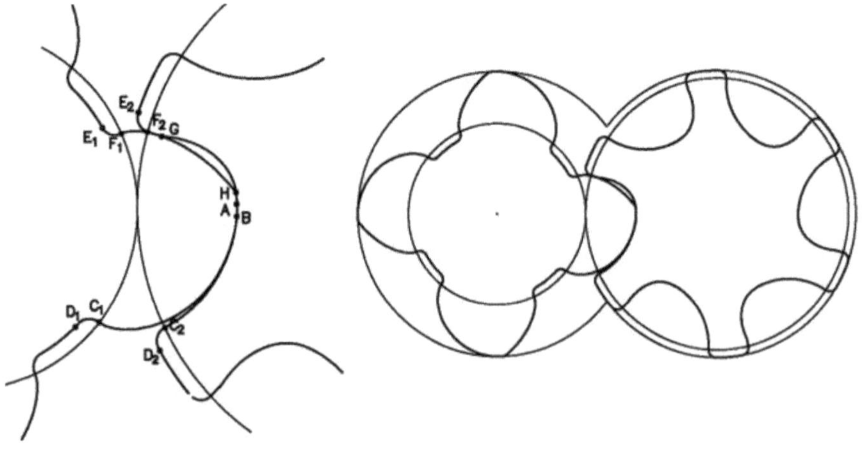

Fig. 2.11. SRM "A" Profile

the opposite lobe side. F_2E_2 is a circle with the centre on the gate pitch and finally, E_2D_2 is a circle with the centre on the gate axis.

More details on the "A" profile are published by Amosov et al., 1977 and by Rinder, 1979.

The "A" profile is a good example of how a good and simple idea evolved into a complicated result. Thus the "A" profile was continuously subjected to changes which resulted in the "C" profile. This was mainly generated to improve the profile manufacturability. Finally, a completely new profile, the "D" profile was generated to introduce a new development in profile gearing and to increase the gate rotor torque.

Despite the complexity of its final form the "A" profile emerged to be the most popular screw compressor profile, especially after its patent expired.

2.4.9 SRM "D" Profile

The SRM "D" profile, shown in Fig. 2.12, is generated exclusively by circles with the centres off the rotor pitch circles.

Similar to the Demonstrator, C_2D_2 is an eccentric circle of radius r_3 on the gate rotor. B_1C_1 is an eccentric circle of radius r_1, which, together with the small circular arc A_1J_1 of radius r_2, is positioned on the main rotor. G_2H_2 is a small circular arc on the gate rotor and E_2F_2 is a circular arc on the gate rotor. F_2G_2 is a relatively large circular arc on the gate rotor which produces a corresponding curve of the smallest possible curvature on the main rotor. Both circular arc, B_2C_2 and F_2G_2 ensure a large radius of curvature in the pitch circle area. This avoids high stresses in the rotor contact region.

Fig. 2.12. SRM "D" Profile

2.4.10 SRM "G" Profile

The "G" profile was introduced by SRM in the late nineteen nineties as a replacement for the "D" rotor and is shown in Fig. 2.13. Compared with the "D" rotor, the "G" rotor has the unique feature of two additional circles in the addendum area on both lobes of the main rotor, close to the pitch circle. This feature improves the rotor contact and, additionally, generates shorter sealing lines. This can be seen in Fig. 2.13, where a rotor featuring "G" profile characteristics only on its flat side through segment H_1I_1 is presented.

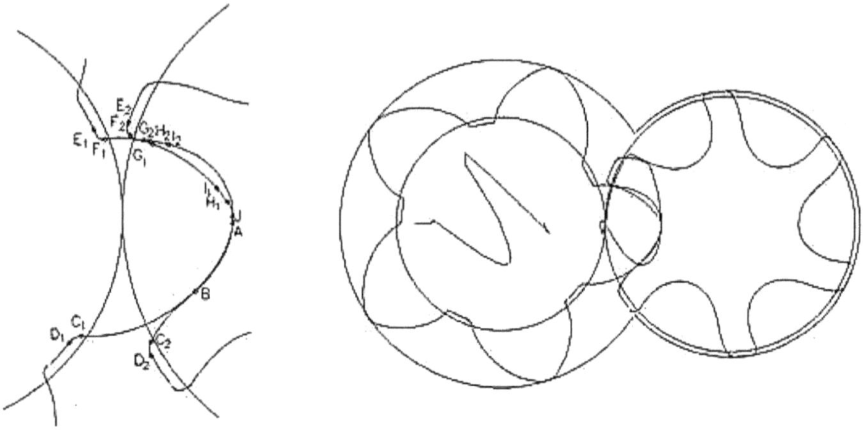

Fig. 2.13. SRM "G" Profile

2.4.11 City "N" Rack Generated Rotor Profile

"N" rotors are calculated by a rack generation procedure. This distinguishes them from any others. In this case, the large blow-hole area, which is a characteristic of rack generated rotors, is overcome by generating the high pressure side of the rack by means of a rotor conjugate procedure. This undercuts the single appropriate curve on the rack. Such a rack is then used for profiling both the main and the gate rotors. The method and its extensions were used by the authors to create a number of different rotor profiles, some of them used by Stosic et al., 1986, and Hanjalic and Stosic, 1994. One of the applications of the rack generation procedure is described in Stosic, 1996.

The following is a brief description of a rack generated "N" rotor profile, typical of a family of rotor profiles designed for the efficient compression of air, common refrigerants and a number of process gases. The rotors are generated by the combined rack-rotor generation procedure whose features are such that it may be readily modified further to optimize performance for any specific application.

The coordinates of all primary arcs on the rack are summarized here relative to the rack coordinate system. The lobe of the rack is divided into several arcs. The divisions between the profile arcs are denoted by capital letters and each arc is defined separately, as shown in the Figs. 2.14 and 2.15 where the rack and the rotors are shown.

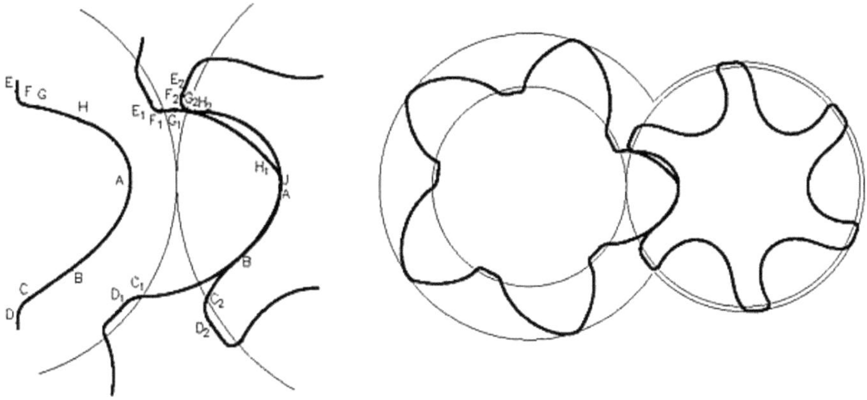

Fig. 2.14. Rack generated "N" Profile

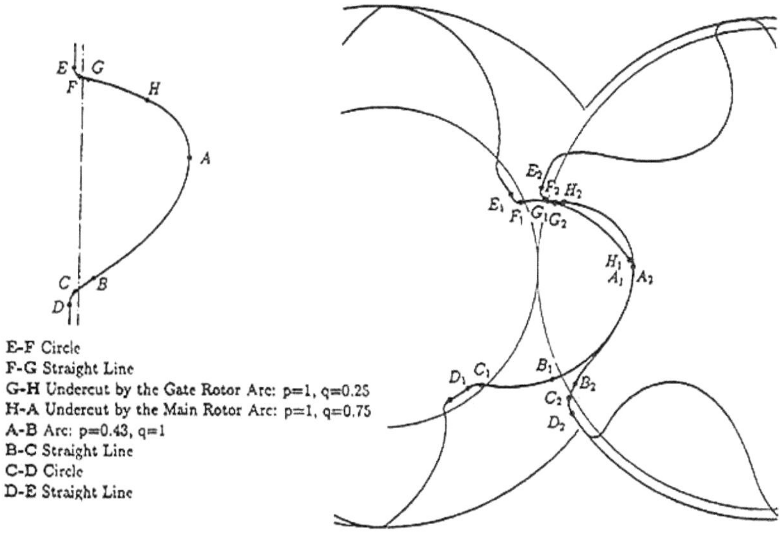

Fig. 2.15. "N" rotor primary curves given on rack

All curves are given as a "general arc" expressed as: $ax^p + by^q = 1$. Thus straight lines, circles, parabolae, ellipses and hyperbolae are all easily described by selecting appropriate values for parameters a, b, p and q.

Segment DE is a straight line on the rack, EF is a circular arc of radius r_4, segment FG is a straight line for the upper involute, $p = q = 1$, while segment GH on the rack is a meshing curve generated by the circular arc G_2H_2 on the gate rotor. Segment HJ on the rack is a meshing curve generated by the circular arc H_1J_1 of radius r_2 on the main rotor. Segment JA is a circular arc of radius r on the rack, AB is an arc which can be either a circle or a parabola, a hyperbola or an ellipse, segment BC is a straight line on the rack matching the involute on the rotor round lobe and CD is a circular arc on the rack, radius r_3.

More details of the "N" profile can be found in Stosic, 1994.

2.4.12 Characteristics of "N" Profile

Sample illustrations of the "N" profile in 2-3, 3-5, 4-5, 4-6, 5-6, 5-7 and 6-7 configurations are given in Figs. 2.16 to Fig. 2.23. It should be noted that all rotors considered were obtained automatically from a computer code by simply specifying the number of lobes in the main and gate rotors, and the lobe curves in the general form.

A variety of modified profiles is possible. The "N" profile design is a compromise between full tightness, small blow-hole area, large displacement, short

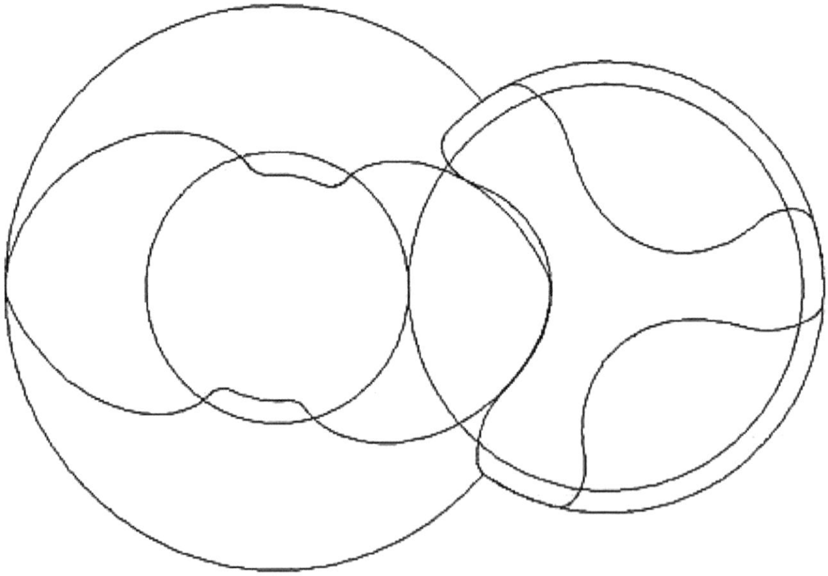

Fig. 2.16. "N" Rotors in 2-3 configuration

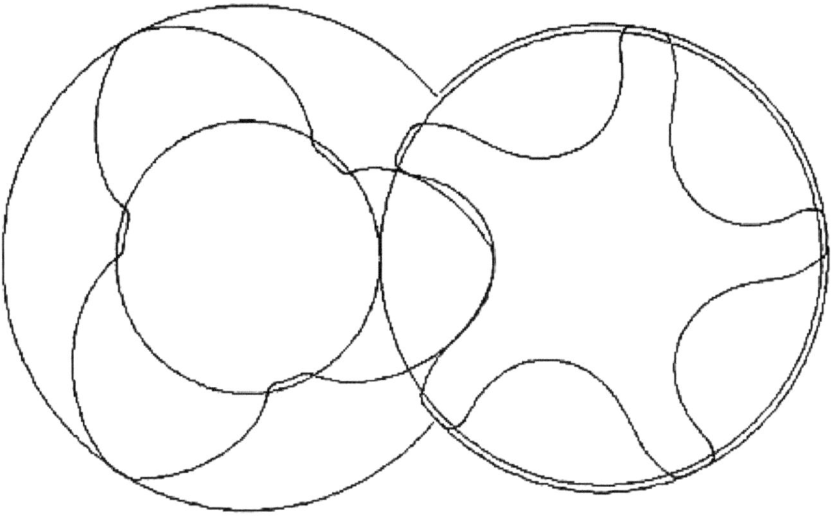

Fig. 2.17. "N" Rotors in 3-5 configuration

Fig. 2.18. "N" Rotors in 4-5 configuration

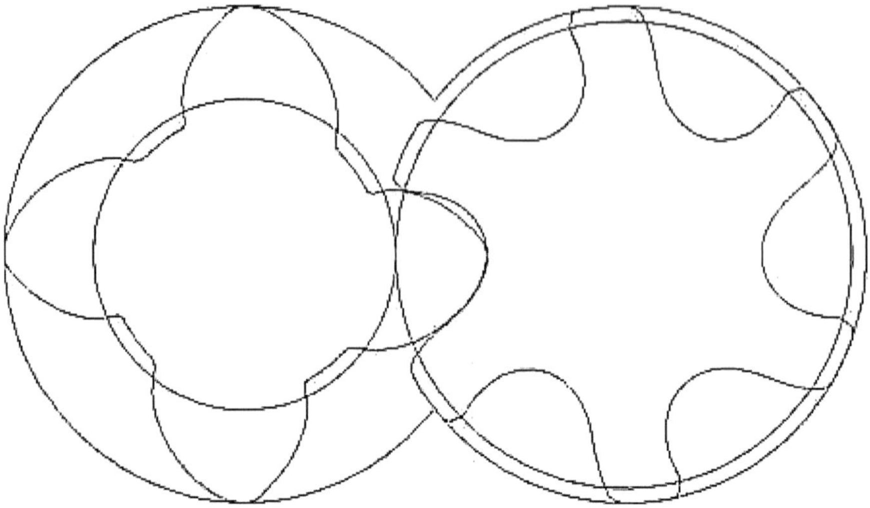

Fig. 2.19. "N" Rotors in 4-6 configuration

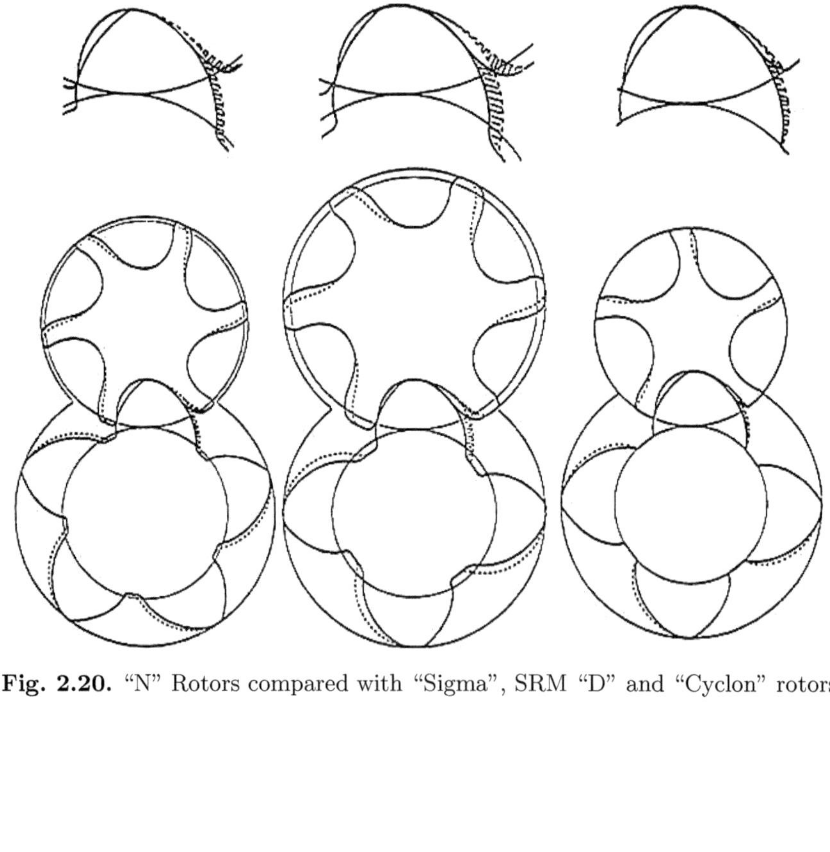

Fig. 2.20. "N" Rotors compared with "Sigma", SRM "D" and "Cyclon" rotors

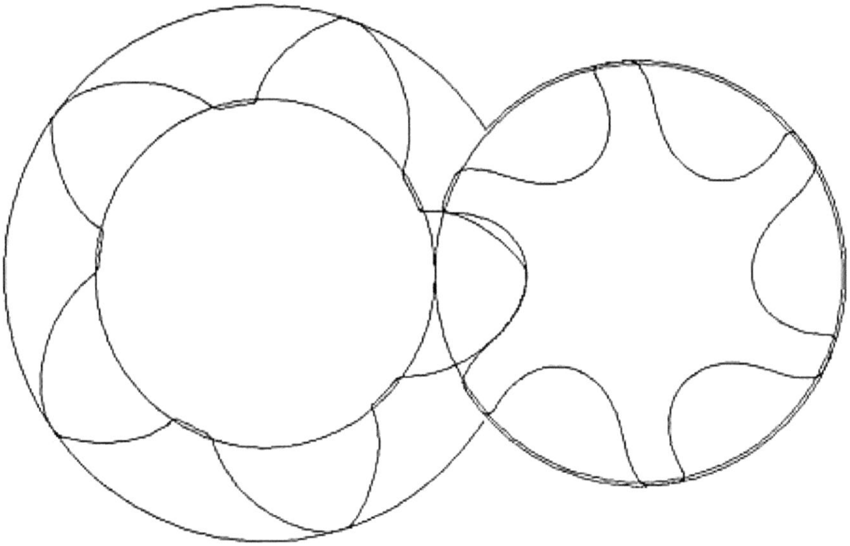

Fig. 2.21. "N" Rotors in 5-6 configuration

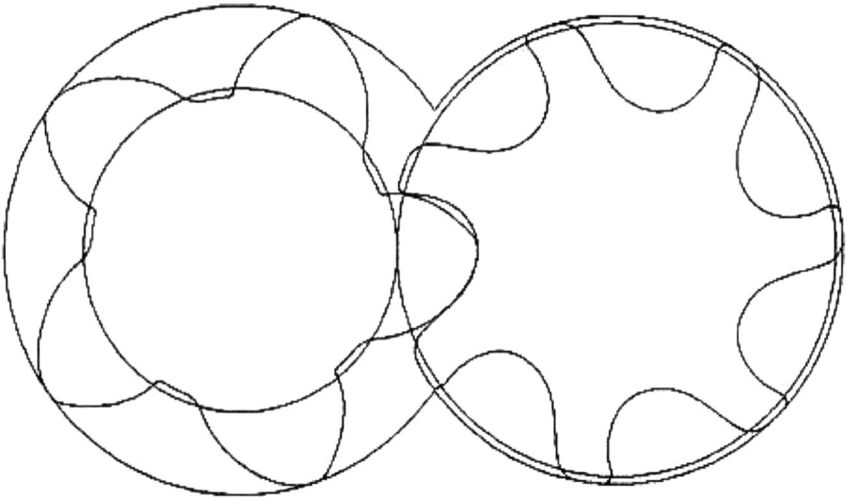

Fig. 2.22. "N" Rotors in 5-7 configuration

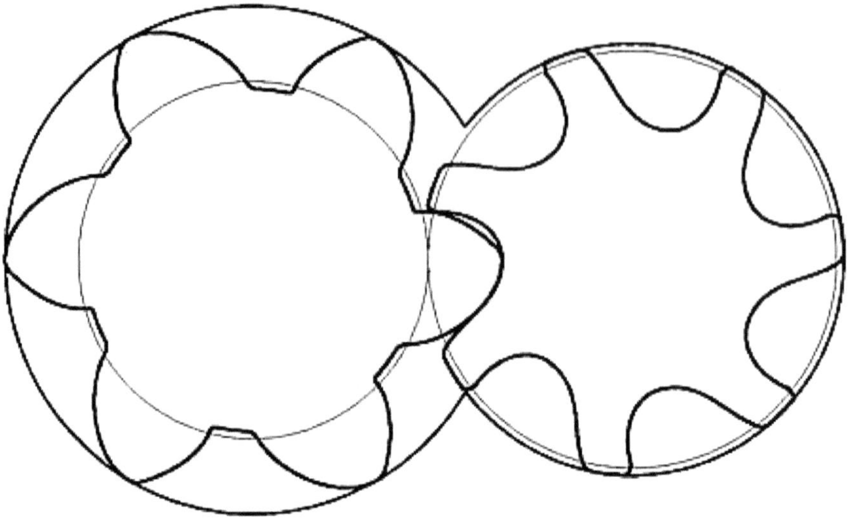

Fig. 2.23. "N" rotors in 6/7 configuration

sealing lines, small confined volumes, involute rotor contact and proper gate rotor torque distribution together with high rotor mechanical rigidity.

The number of lobes required varies according to the designated compressor duty. The 3/5 arrangement is most suited for dry air compression, the 4/5 and 5/6 for oil flooded compressors with a moderate pressure difference and the 6/7 for high pressure and large built-in volume ratio refrigeration applications.

Although the full evaluation of a rotor profile requires more than just a geometric assessment, some of the key features of the "N" profile may be readily appreciated by comparing it with three of the most popular screw rotor profiles already described here, (a) The "Sigma" profile by Bammert, 1979, (b) the SRM "D" profile by Astberg 1982, and (c) the "Cyclon" profile by Hough and Morris, 1984. All these rotors are shown in Fig. 2.20 where it can be seen that the "N" profiles have a greater throughput and a stiffer gate rotor for all cases when other characteristics such as the blow-hole area, confined volume and high pressure sealing line lengths are identical.

Also, the low pressure sealing lines are shorter, but this is less important because the corresponding clearance can be kept small.

The blow-hole area may be controlled by adjustment of the tip radii on both the main and gate rotors and also by making the gate outer diameter equal to or less than the pitch diameter. Also the sealing lines can be kept very short by constructing most of the rotor profile from circles whose centres are close to the pitch circle. But, any decrease in the blow-hole area will increase the length of the sealing line on the flat rotor side. A compromise between these trends is therefore required to obtain the best result.

Rotor instability is often caused by the torque distribution in the gate rotor changing direction during a complete cycle. The profile generation procedure described in this paper makes it possible to control the torque on the gate rotor and thus avoid such effects. Furthermore, full involute contact between the "N" rotors enables any additional contact load to be absorbed more easily than with any other type of rotor. Two rotor pairs are shown in Fig. 2.24 the first exhibits what is described as "negative" gate rotor torque while the second shows the more usual "positive" torque.

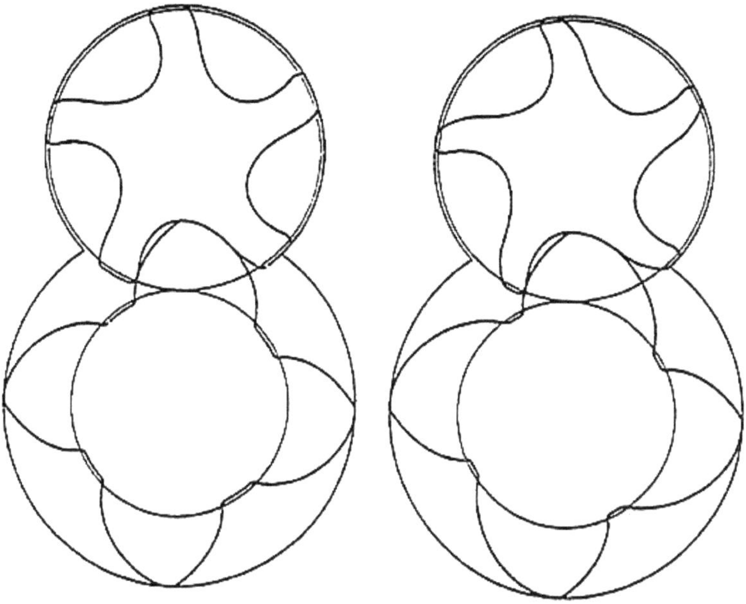

Fig. 2.24. "N" with negative torque, *left* and positive torque, *right*

2.4.13 Blower Rotor Profile

The blower profile, shown in Fig. 2.25 is symmetrical. Therefore only one quarter of it needs to be specified in order to define the whole rotor. It consists of two segments, a very small circle on the rotor lobe tip and a straight line. The circle slides and generates cycloids, while the straight line generates involutes.

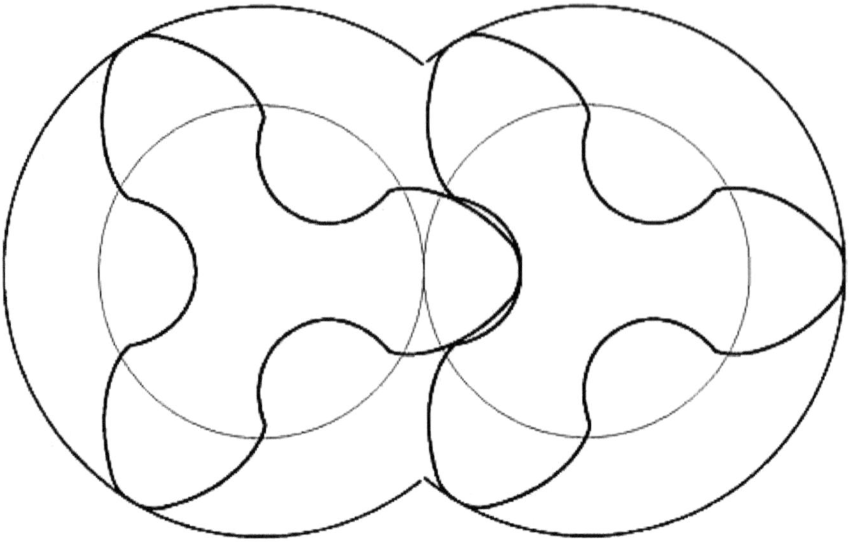

Fig. 2.25. Blower profile

2.5 Identification of Rotor Position in Compressor Bearings

The rotor axial and radial forces are transferred to the housing by the bearings. Rolling element bearings are normally chosen for small and medium screw compressors and these must be carefully selected to obtain a satisfactory design. Usually, two bearings are employed on the discharge end of each of the rotor shafts in order to absorb the radial and axial loads separately. Also, the distance between the rotor centre lines is in part determined by the bearing size and internal clearance. Any manufacturing imperfection in the bearing housing, like displacement or eccentricity, will change the rotor position and thereby influence the compressor behaviour.

The system of rotors in screw compressor bearings is presented in Fig. 2.26. The rotor shafts are parallel and their positions are defined by axes Z_1 and Z_2.

The bearings are labelled 1 to 4, and their clearances, as well as the manufacturing tolerances of the bearing bores, δ_x and δ_y in the x and y directions respectively, are presented in the same figure. The rotor centre distance is C and the axial span between the bearings is a.

All imperfections in the manufacture of screw compressor rotors should fall within and be accounted for by production tolerances. These are the wrong position of the bearing bores, eccentricity of the rotor shafts, bearing clearances and imperfections and rotor misalignment. Together, they account for the rotor shafts not being parallel. Let rotor movement δ_y in the y direction contain all displacements, which are presented in Fig. 2.27, and cause virtual rotation of the rotors around the X_1, and X_2 axes, as shown in Fig. 2.27. Let

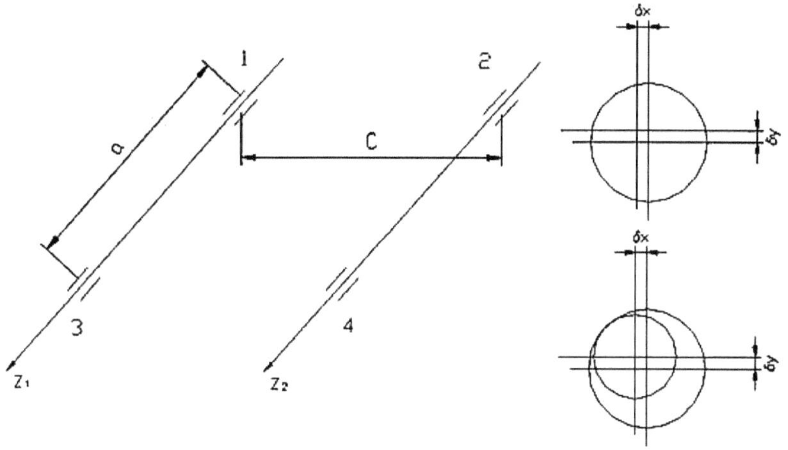

Fig. 2.26. Rotor shafts in the compressor housing and displacement in bearings

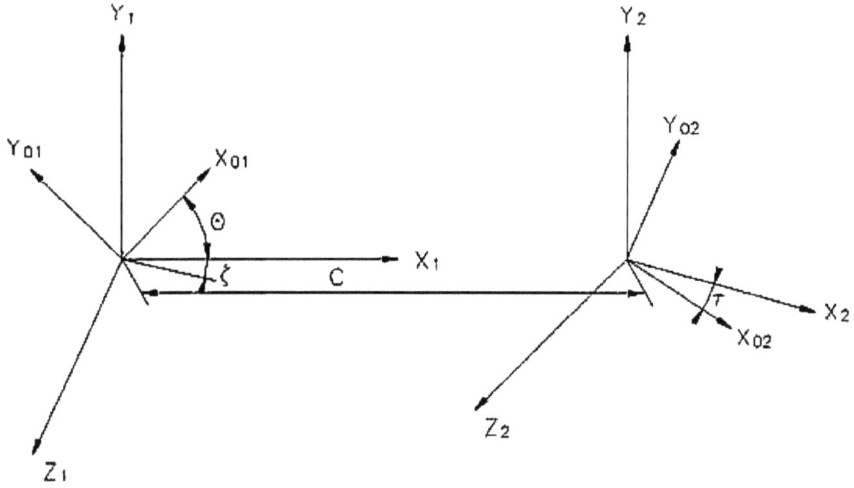

Fig. 2.27. Rotors with intersecting shafts and their coordinate systems

rotor movement δ_x in the x direction cause rotation around the Y_1, and Y_2 axes, as shown in Fig. 2.28. The movement δ_x can cause the rotor shafts to intersect. However, the movement δ_y causes the shafts to become non-parallel and non-intersecting. These both change the nature of the rotor position so that the shafts can no longer be regarded as parallel. The following analytical approach enables the rotor movement to be calculated and accounts for these changes.

Vectors $\mathbf{r}_1 = [x_1, y_1, z_1]$ and $\mathbf{r}_2$, now represent the helicoid surfaces of the main and gate rotors on intersecting shafts. The shaft angle ζ, is the rotation about Y.

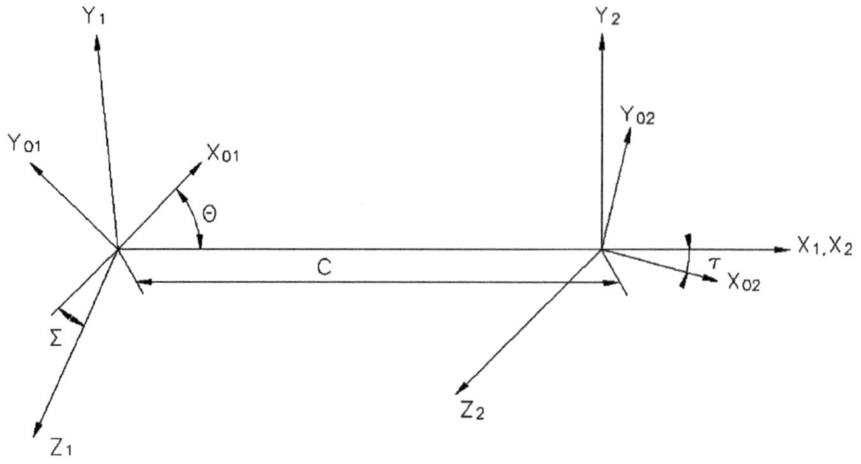

Fig. 2.28. Rotors with non-parallel and non-intersecting shafts and their coordinate systems

$$\mathbf{r}_2 = [x_2, y_2, z_2] = [x_1 \cos \varsigma - z_1 \sin \varsigma - C, y_1, x_1 \sin \varsigma + z_1 \cos \varsigma] \tag{2.15}$$

$$tg\varsigma = \frac{\delta_x}{a} \tag{2.16}$$

Since this rotation angle is usually very small, the relationship (2.16) can be assumed. Equation (2.15) can then be simplified for further analysis.

$$\mathbf{r}_2 = [x_2, y_2, z_2] = [x_1 - z_1\varsigma - C, y_1, x_1\varsigma + z_1]$$

The rotation ζ will result in a displacement $-z_1\zeta$ in the x direction and a displacement $x_1\zeta$ in the z direction, while there is no displacement in the y direction. The displacement vector becomes:

$$\Delta\mathbf{r}_2 = [-z_1\varsigma, 0, x_1\varsigma]$$

In the majority of practical cases, x_1 is small compared with z_1 and only displacement in the x direction need be considered. This means that rotation around the Y axis will, effectively, only change the rotor centre distance. Displacement in the z direction may be significant for the dynamic behaviour of the rotors. Displacement in the z direction will be adjusted by the rotor relative rotation around the Z axis, which can be accompanied by significant angular acceleration. This may cause the rotors to lose contact at certain stages of the compressor cycle and thus create rattling, which may increase the compressor noise.

Since the rotation angle ζ, caused by displacement within the tolerance limits, is very small, a two-dimensional analysis in the rotor end plane can be applied, as is done in the next section.

As shown in Fig. 2.28, where the rotors on the nonparallel and nonintersecting axes are presented, vectors $\mathbf{r}_1 = [x_1, y_1, z_1]$ and $\mathbf{r}_2$, given by (2.10) now represent the helicoid surfaces of the main and gate rotors on the intersecting shafts. Σ is the rotation angle around the X axes given by (2.11).

$$\mathbf{r}_2 = [x_2, y_2, z_2] = [x_1 - C,\ y_1 \cos\Sigma - z_1 \sin\Sigma,\ y_1 \sin\Sigma + z_1 \cos\Sigma] \tag{2.17}$$

$$tg\Sigma = \frac{\delta_y}{a} \tag{2.18}$$

Since angle Σ is very small, it can be expressed in simplified form as in (2.18). Further analysis is then facilitated by writing (2.17) as:

$$\mathbf{r}_2 = [x_2, y_2, z_2] = [x_1 - C, y_1 - z_1\Sigma,\ y_1\Sigma + z_1]$$

The rotation Σ will result in displacement $-z_1\Sigma$ in the y direction and displacement $y_1\Sigma$ in the z direction, while there is no displacement in the x direction. The displacement vector can be written as:

$$\Delta\mathbf{r}_2 = [0, -z_1\Sigma,\ y_1\Sigma]$$

Although, in the majority of practical cases, displacement in the z direction is very small and therefore unimportant for consideration of rotor interference, it may play a role in the dynamic behaviour of the rotors. The displacement in the z direction will be fully compensated by regular rotation of the rotors around the Z axis. However, the angular acceleration involved in this process may cause the rotors to lose contact at some stages of the compressor cycle.

Rotation about the X axis is effectively the same as if the main or gate rotor rotated relatively through angles $\theta = -z_1\Sigma/r_{1w}$ or $\tau = z_1\Sigma/r_{2w}$ respectively and the rotor backlash will be reduced by $z_1\Sigma$. Such an approach substantially simplifies the analysis and allows the problem to be presented in two dimensions in the rotor end plane.

Although the rotor movements, described here are entirely three-dimensional, their two-dimensional presentation in the rotor end plane section can be used for analysis.

Equation (2.2) serves to calculate both the coordinates of the rotor meshing points x_2, y_2 on the rotor helicoids and x_{02}, y_{02} in the end plane from the given rotor coordinates points x_{01} and y_{01}. It may also be used to determine the contact line coordinates and paths of contact between the rotors. The sealing line of screw compressor rotors is somewhat similar to the rotor contact line. Since there is a clearance gap between rotors, sealing is effected at the points of the most proximate rotor position. A convenient practice to obtain the clearance gap between the rotors is to consider the gap as the shortest distance between the rotors in a section normal to the rotor helicoids. The end plane clearance gap can then be obtained from the normal clearance by appropriate transformation.

If δ is the normal clearance between the rotor helicoid surfaces, the cross product of the $\mathbf{r}$ derivatives, given in the left hand side of (2.5), which defines the direction normal to the helicoids, can be used to calculate the coordinates of the rotor helicoids x_n and y_n from x and y to which the clearance is added as:

$$x_n = x + p\frac{\delta}{D}\frac{dy}{dt}, \quad y_n = y - p\frac{\delta}{D}\frac{dx}{dt}, \quad z_n = \frac{\delta}{D}\left(x\frac{dx}{dt} + y\frac{dy}{dt}\right) \tag{2.19}$$

where the denominator D is given as:

$$D = \sqrt{p^2\left(\frac{dx}{dt}\right)^2 + p^2\left(\frac{dy}{dt}\right)^2 + \left(x\frac{dx}{dt} + y\frac{dy}{dt}\right)^2} \tag{2.20}$$

x_n and y_n serve to calculate new rotor end plane coordinates, x_{0n} and y_{0n}, with the clearances obtained for angles $\theta = z_n/p$ and τ respectively. These x_{0n} and y_{0n} now serve to calculate the transverse clearance δ_0 as the difference between them, as well as the original rotor coordinates x_0 and y_0.

If by any means, the rotors change their relative position, the clearance distribution at one end of the rotors may be reduced to zero on the flat side of the rotor lobes. In such a case, rotor contact will be prohibitively long on the flat side of the profile, where the dominant relative rotor motion is sliding, as shown in Fig. 2.29. This indicates that rotor seizure will almost certainly occur in that region if the rotors come into contact with each other.

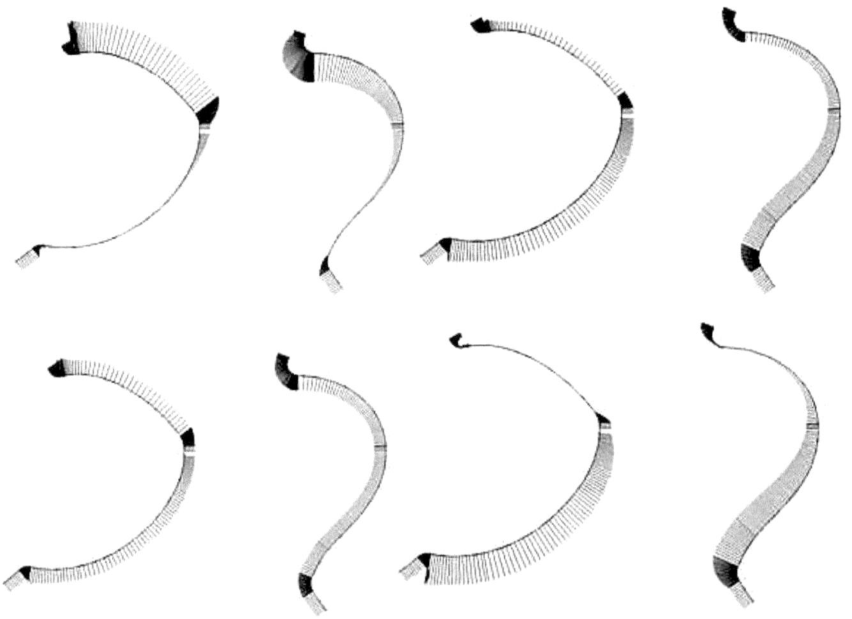

Fig. 2.29. Clearance distribution between the rotors: at suction, mid rotors, and discharge with possible rotor contact at the discharge

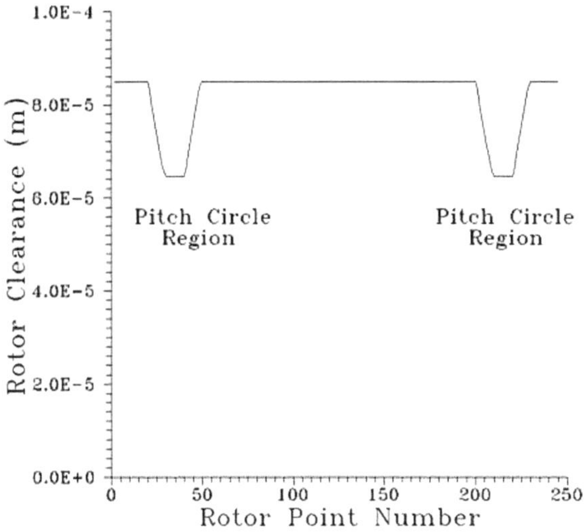

Fig. 2.30. Variable clearance distribution applied to the rotors

It follows that the clearance distribution should be non-uniform to avoid hard rotor contact in rotor areas where sliding motion between the rotors is dominant.

In Fig. 2.30, a reduced clearance of 65 µm is presented, which is now applied in rotor regions close to the rotor pitch circles, while in other regions it is kept at 85 µm, as was done by Edstroem, 1992. As can be seen in Fig. 2.31, the situation regarding rotor contact is now quite different. This is maintained along the rotor contact belt close to the rotor pitch circles and fully avoided at other locations. It follows that if contact occurred, it would be of a rolling character rather than a combination of rolling and sliding or even pure sliding. Such contact will not generate excessive heat and could therefore be maintained for a longer period without damaging the rotors until contact ceases or the compressor is stopped.

2.6 Tools for Rotor Manufacture

This section describes the generation of formed tools for screw compressor hobbing, milling and grinding based on the envelope gearing procedure.

2.6.1 Hobbing Tools

A screw compressor rotor and its formed hobbing tool are equivalent to a pair of meshing crossed helical gears with nonparallel and nonintersecting axes. Their general meshing condition is given in Appendix A. Apart from the gashes forming the cutter faces, the hob is simply a helical gear in which

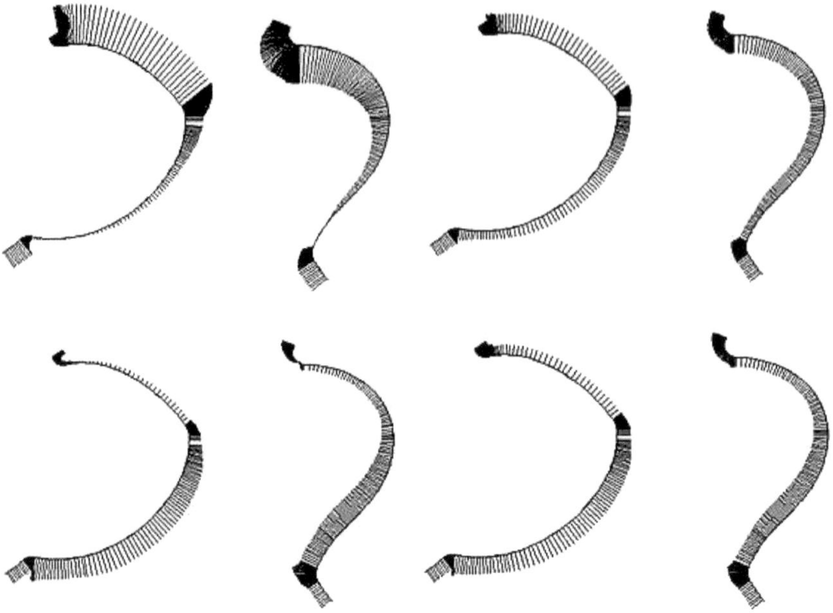

Fig. 2.31. Clearance distribution between the rotors: at suction, mid of rotor and discharge with a possible rotor contact at the discharge

each tooth is referred to as a thread, Colburne, 1987. Owing to their axes not being parallel, there is only point contact between them whereas there is line contact between the screw machine rotors. The need to satisfy the meshing equation given in Appendix A, leads to the rotor – hob meshing requirement for the given rotor transverse coordinate points x_{01} and y_{01} and their first derivative $\frac{dy_{01}}{dx_{01}}$. The hob transverse coordinate points x_{02} and y_{02} can then be calculated. These are sufficient to obtain the coordinate $R_2 = \sqrt{x_{01}^2 + y_{01}^2}$. The axial coordinate z_2, calculated directly, and R_2 are hob axial plane coordinates which define the hob geometry.

The transverse coordinates of the screw machine rotors, described in the previous section, are used as an example here to produce hob coordinates. The rotor unit leads p_1 are 48.754 mm for the main and −58.504 mm for the gate rotor. Single lobe hobs are generated for unit leads p_2: 6.291 mm for the main rotor and −6.291 mm for the gate rotor. The corresponding hob helix angles ψ are 85° and 95°. The same rotor-to-hob centre distance $C = 110$ mm and the shaft angle $\Sigma = 50°$ are given for both rotors. Figure 2.32 contains a view to the hob.

Reverse calculation of the hob – screw rotor transformation, also given in Appendix A, permits the determination of the transverse rotor profile coordinates which will be obtained as a result of the manufacturing process. These may be compared with those originally specified to determine the effect of

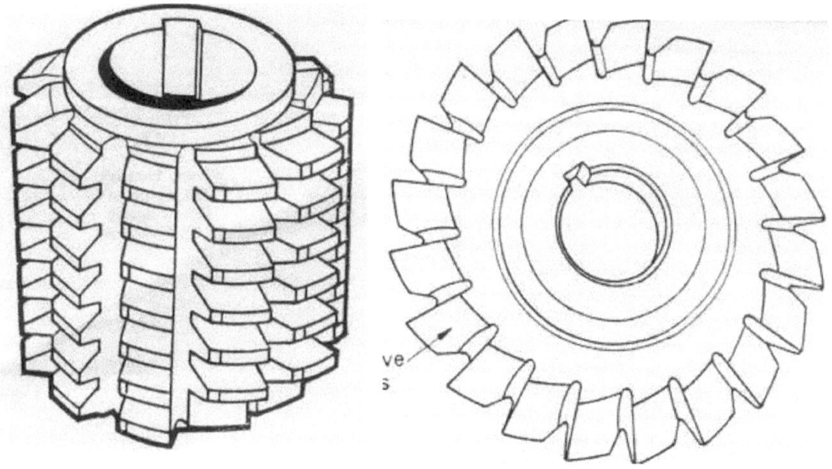

Fig. 2.32. Rotor manufacturing: hobbing tool *left*, *right* milling tool

manufacturing errors such as imperfect tool setting or tool and rotor deformation upon the final rotor profile.

For the purpose of reverse transformation, the hob longitudinal plane coordinates R_2 and z_2 and $\frac{dR_2}{dz_2}$ should be given. The axial coordinate z_2 is used to calculate $\tau = z_2/p_2$, which is then used to calculate the hob transverse coordinates:

$$x_{02} = R_2 \cos\tau, \qquad y_{02} = R_2 \sin\tau \tag{2.21}$$

These are then used as the given coordinates to produce a meshing criterion and the transverse plane coordinates of the "manufactured" rotors.

A comparison between the original rotors and the manufactured rotors is given in Fig. 2.33 with the difference between them scaled 100 times. Two types of error are considered. The left gate rotor, is produced with 30 μm offset in the centre distance between the rotor and the tool, and the main rotor with

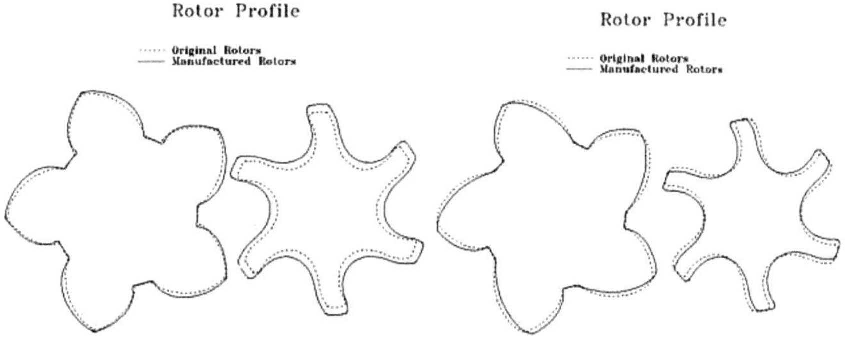

Fig. 2.33. Manufacturing imperfections

0.2° offset in the tool shaft angle Σ. Details of this particular meshing method are given by Stosic 1998.

2.6.2 Milling and Grinding Tools

Formed milling and grinding tools may also be generated by placing $p_2 = 0$ in the general meshing equation, given in Appendix A, and then following the procedure of this section. The resulting meshing condition now reads as:

$$[C - x_1 + p_1 \cot \Sigma] \left(x_1 \frac{\partial x_1}{\partial t} + y_1 \frac{\partial y_1}{\partial t} \right) + p_1 \left[p_1 \theta \frac{\partial y_1}{\partial t} - C \cot \Sigma \frac{\partial x_1}{\partial t} \right] = 0 \tag{2.22}$$

However in this case, when one expects to obtain screw rotor coordinates from the tool coordinates, the singularity imposed does not permit the calculation of the tool transverse plane coordinates. The main meshing condition cannot therefore be applied. For this purpose another condition is derived for the reverse milling tool to rotor transformation from which the meshing angle τ is calculated:

$$\left(R_2 + z_2 \frac{dz_2}{dR_2} \right) \cos \tau + (p_1 + C \cot \Sigma) \frac{dz_2}{dR_2} \sin \tau + p_1 \cot \Sigma - C = 0 \tag{2.23}$$

Once obtained, τ will serve to calculate the rotor coordinates after the "manufacturing" process. The obtained rotor coordinates will contain all manufacturing imperfections, like mismatch of the rotor – tool centre distance, error in the rotor – tool shaft angle, axial shift of the tool or tool deformation during the process as they are input to the calculation process. A full account of this useful procedure is given by Stosic 1998.

2.6.3 Quantification of Manufacturing Imperfections

The rotor – tool transformation is used here for milling tool profile generation. The reverse procedure is used to calculate the "manufactured" rotors. The rack generated 5-6 128 mm rotors described by Stosic, 1997a are used as given profiles: $x(t)$ and $y(t)$. Then a tool – rotor transformation is used to quantify the influence of manufacturing imperfections upon the quality of the produced rotor profile. Both, linear and angular offset were considered.

Figure 2.33 presents the rotors, the main manufactured with the shaft angle offset 0.5° and the gate with the centre distance offset 40 µm from that of the original rotors given by the dashed line on the left. On the right, the rotors are manufactured with imperfections, the main with a tool axial offset of 40 µm and the gate with a certain tool body deformation which resulted in 0.5° offset of the relative motion angle θ. The original rotors are given by the dashed line.

3

Calculation of Screw Compressor Performance

Screw compressor performance is governed by the interactive effects of thermodynamic and fluid flow processes and the machine geometry and thus can be calculated reliably only by their simultaneous consideration. This may be achieved by mathematical modelling in one or more dimensions. For most applications, a one dimensional model is sufficient and this is described in full. 3-D modelling is more complex and is presented here only in outline. A more detailed presentation of this will be made in a separate publication.

3.1 One Dimensional Mathematical Model

The algorithm used to describe the thermodynamic and fluid flow processes in a screw compressor is based on a mathematical model. This defines the instantaneous volume of the working chamber and its change with rotational angle or time, to which the conservation equations of energy and mass continuity are applied, together with a set of algebraic relationships used to define various phenomena related to the suction, compression and discharge of the working fluid. These form a set of simultaneous non-linear differential equations which cannot be solved in closed form.

The solution of the equation set is performed numerically by means of the Runge-Kutta 4th order method, with appropriate initial and boundary conditions.

The model accounts for a number of "real-life" effects, which may significantly influence the performance of a real compressor. These make it suitable for a wide range of applications and include the following:

- The working fluid compressed can be any gas or liquid-gas mixture for which an equation of state and internal energy-enthalpy relation is known, i.e. any ideal or real gas or liquid-gas mixture of known properties.
- The model accounts for heat transfer between the gas and the compressor rotors or its casing in a form, which though approximate, reproduces the overall effect to a good first order level of accuracy.

- The model accounts for leakage of the working medium through the clearances between the two rotors and between the rotors and the stationary parts of the compressor.
- The process equations and the subroutines for their solution are independent of those which define the compressor geometry. Hence, the model can be readily adapted to estimate the performance of any geometry or type of positive displacement machine.
- The effects of liquid injection, including that of oil, water, or refrigerant can be accounted for during the suction, compression and discharge stages.
- A set of subroutines to estimate the thermodynamic properties and changes of state of the working fluid during the entire compressor cycle of operations completes the equation set and thereby enables it to be solved.

Certain assumptions had to be introduced to ensure efficient computation. These do not impose any limitations on the model nor cause significant departures from the real processes and are as follows:

- The fluid flow in the model is assumed to be quasi one-dimensional
- Kinetic energy changes of the working fluid within the working chamber are negligible compared to internal energy changes.
- Gas or gas-liquid inflow to and outflow from the compressor ports is assumed to be isentropic.
- Leakage flow of the fluid through the clearances is assumed to be adiabatic.

3.1.1 Conservation Equations for Control Volume and Auxiliary Relationships

The working chamber of a screw machine is the space within it that contains the working fluid. This is a typical example of an open thermodynamic system in which the mass flow varies with time. This, as well as the suction and discharge plenums, can be defined by a control volume for which the differential equations of the conservation laws for energy and mass are written. These are derived in Appendix B, using Reynolds Transport Theorem.

A feature of the model is the use of the non-steady flow energy equation to compute the thermodynamic and flow processes in a screw machine in terms of rotational angle or time and how these are affected by rotor profile modifications. Internal energy, rather than enthalpy, is then the derived variable. This is computationally more convenient than using enthalpy as the derived variable since, even in the case of real fluids, it may be derived, without reference to pressure. Computation is then carried out through a series of iterative cycles until the solution converges. Pressure, which is the desired output variable, can then be derived directly from it, together with the remaining required thermodynamic properties.

The following forms of the conservation equations have been employed in the model:

The Conservation of Internal Energy

$$\omega\left(\frac{dU}{d\theta}\right) = \dot{m}_{\text{in}}h_{\text{in}} - \dot{m}_{\text{out}}h_{\text{out}} + Q - \omega p\frac{dV}{d\theta} \tag{3.1}$$

where θ is angle of rotation of the main rotor, $h = h(\theta)$ is specific enthalpy, $\dot{m} = \dot{m}(\theta)$ is mass flow rate $p = p(\theta)$, fluid pressure in the working chamber control volume, $\dot{Q} = \dot{Q}(\theta)$, heat transfer between the fluid and the compressor surrounding, $\dot{V} = \dot{V}(\theta)$ local volume of the compressor working chamber.

In the above equation the subscripts in and out denote the fluid inflow and outflow.

The fluid total enthalpy inflow consists of the following components:

$$\dot{m}_{\text{in}}h_{\text{in}} = \dot{m}_{\text{suc}}h_{\text{suc}} + \dot{m}_{l,g}h_{l,g} + \dot{m}_{\text{oil}}h_{\text{oil}} \tag{3.2}$$

where subscripts l, g denote leakage gain suc, suction conditions, and oil denotes oil.

The fluid total outflow enthalpy consists of:

$$\dot{m}_{\text{out}}h_{\text{out}} = \dot{m}_{\text{dis}}h_{\text{dis}} + \dot{m}_{l,l}h_{l,l} \tag{3.3}$$

where indices l, l denote leakage loss and dis denotes the discharge conditions with $\dot{m}_{\text{dis}}$ denoting the discharge mass flow rate of the gas contaminated with the oil or other liquid injected.

The right hand side of the energy equation consists of the following terms which are modelled:

- The heat exchange between the fluid and the compressor screw rotors and casing and through them to the surrounding, due to the difference in temperatures of gas and the casing and rotor surfaces is accounted for by the heat transfer coefficient evaluated from the expression $Nu = 0.023\ \text{Re}^{0.8}$. For the characteristic length in the Reynolds and Nusselt number the difference between the outer and inner diameters of the main rotor was adopted. This may not be the most appropriate dimension for this purpose, but the characteristic length appears in the expression for the heat transfer coefficient with the exponent of 0.2 and therefore has little influence as long as it remains within the same order of magnitude as other characteristic dimensions of the machine and as long as it characterizes the compressor size. The characteristic velocity for the Re number is computed from the local mass flow and the cross-sectional area. Here the surface over which the heat is exchanged, as well as the wall temperature, depend on the rotation angle θ of the main rotor.
- The energy gain due to the gas inflow into the working volume is represented by the product of the mass intake and its averaged enthalpy. As such, the energy inflow varies with the rotational angle. During the suction period, gas enters the working volume bringing the averaged gas enthalpy,

which dominates in the suction chamber. However, during the time when the suction port is closed, a certain amount of the compressed gas leaks into the compressor working chamber through the clearances. The mass of this gas, as well as its enthalpy are determined on the basis of the gas leakage equations. The working volume is filled with gas due to leakage only when the gas pressure in the space around the working volume is higher, otherwise there is no leakage, or it is in the opposite direction, i.e. from the working chamber towards other plenums.

- The total inflow enthalpy is further corrected by the amount of enthalpy brought into the working chamber by the injected oil.
- The energy loss due to the gas outflow from the working volume is defined by the product of the mass outflow and its averaged gas enthalpy. During delivery, this is the compressed gas entering the discharge plenum, while, in the case of expansion due to inappropriate discharge pressure, this is the gas which leaks through the clearances from the working volume into the neighbouring space at a lower pressure. If the pressure in the working chamber is lower than that in the discharge chamber and if the discharge port is open, the flow will be in the reverse direction, i.e. from the discharge plenum into the working chamber. The change of mass has a negative sign and its assumed enthalpy is equal to the averaged gas enthalpy in the pressure chamber.
- The thermodynamic work supplied to the gas during the compression process is represented by the term $p\frac{dV}{d\theta}$. This term is evaluated from the local pressure and local volume change rate. The latter is obtained from the relationships defining the screw kinematics which yield the instantaneous working volume and its change with rotation angle. In fact the term $dV/d\varphi$ can be identified with the instantaneous interlobe area, corrected for the captured and overlapping areas.
- If oil or other fluid is injected into the working chamber of the compressor, the oil mass inflow and its enthalpy should be included in the inflow terms. In spite of the fact that the oil mass fraction in the mixture is significant, its effect upon the volume flow rate is only marginal because the oil volume fraction is usually very small. The total fluid mass outflow also includes the injected oil, the greater part of which remains mixed with the working fluid. Heat transfer between the gas and oil droplets is described by a first order differential equation.

The Mass Continuity Equation

$$\omega\frac{d\dot{m}}{d\theta} = \dot{m}_{\text{in}}h_{\text{in}} - \dot{m}_{\text{out}}h_{\text{out}} \tag{3.4}$$

The mass inflow rate consists of:

$$\dot{m}_{\text{in}}h_{\text{in}} = \dot{m}_{\text{suc}} + \dot{m}_{l,g} + \dot{m}_{\text{oil}} \tag{3.5}$$

The mass outflow rate consists of:

$$\dot{m}_{\text{out}} h_{\text{out}} = \dot{m}_{\text{dis}} + \dot{m}_{l,l} \tag{3.6}$$

Each of the mass flow rate satisfies the continuity equation

$$\dot{m} = \rho w A \tag{3.7}$$

where w[m/s] denotes fluid velocity, ρ – fluid density and A – the flow cross-section area.

The instantaneous density $\rho = \rho(\theta)$ is obtained from the instantaneous mass m trapped in the control volume and the size of the corresponding instantaneous volume V, as $\rho = m/V$.

3.1.2 Suction and Discharge Ports

The cross-section area A is obtained from the compressor geometry and it may be considered as a periodic function of the angle of rotation θ. The suction port area is defined by:

$$A_{\text{suc}} = A_{\text{suc},0} \sin\left(\pi \frac{\theta}{\theta_{\text{suc}}}\right) \tag{3.8}$$

where suc means the starting value of θ at the moment of the suction port opening, and $A_{\text{suc},0}$ denotes the maximum value of the suction port cross-section area. The reference value of the rotation angle θ is assumed at the suction port closing so that suction ends at $\theta = 0$, if not specified differently.

The discharge port area is likewise defined by:

$$A_{\text{dis}} = A_{\text{dis},0} \sin\left(\pi \frac{\theta - \theta_c}{\theta_e - \theta_s}\right) \tag{3.9}$$

where subscript e denotes the end of discharge, c denotes the end of compression and $A_{\text{dis},0}$ stands for the maximum value of the discharge port cross-sectional area.

Suction and Discharge Port Fluid Velocities

$$w = \mu\sqrt{2\left(h_2 - h_1\right)} \tag{3.10}$$

where μ is the suction/discharge orifice flow coefficient, while subscripts 1 and 2 denote the conditions downstream and upstream of the considered port. The provision supplied in the computer code will calculate for a reverse flow if $h_2 < h_1$.

3.1.3 Gas Leakages

Leakages in a screw machine amount to a substantial part of the total flow rate and therefore play an important role because they influence the process both by affecting the compressor mass flow rate or compressor delivery, i.e. volumetric efficiency and the thermodynamic efficiency of the compression work. For practical computation of the effects of leakage upon the compressor process, it is convenient to distinguish two types of leakages, according to their direction with regard to the working chamber: gain and loss leakages. The gain leakages come from the discharge plenum and from the neighbouring working chamber which has a higher pressure. The loss leakages leave the chamber towards the suction plenum and to the neighbouring chamber with a lower pressure.

Computation of the leakage velocity follows from consideration of the fluid flow through the clearance. The process is essentially adiabatic Fanno-flow. In order to simplify the computation, the flow is is sometimes assumed to be at constant temperature rather than at constant enthalpy. This departure from the prevailing adiabatic conditions has only a marginal influence if the analysis is carried out in differential form, i.e. for the small changes of the rotational angle, as followed in the present model. The present model treats only gas leakage. No attempt is made to account for leakage of a gas-liquid mixture, while the effect of the oil film can be incorporated by an appropriate reduction of the clearance gaps.

An idealized clearance gap is assumed to have a rectangular shape and the mass flow of leaking fluid is expressed by the continuity equation:

$$\dot{m}_l = \mu_l \rho_l w_l A_g \tag{3.11}$$

where r and w are density and velocity of the leaking gas, $A_g = l_g \delta_g$ the clearance gap cross-sectional area, l_g leakage clearance length, sealing line, δ_g leakage clearance width or gap, $\mu = \mu(\mathrm{Re}, \mathrm{Ma})$ the leakage flow discharge coefficient.

Four different sealing lines are distinguished in a screw compressor: the leading tip sealing line formed between the main and gate rotor forward tip and casing, the trailing tip sealing line formed between the main and gate reverse tip and casing, the front sealing line between the discharge rotor front and the housing and the interlobe sealing line between the rotors.

All sealing lines have clearance gaps which form leakage areas. Additionally, the tip leakage areas are accompanied by blow-hole areas.

According to the type and position of leakage clearances, five different leakages can be identified, namely: losses through the trailing tip sealing and front sealing and gains through the leading and front sealing. The fifth, "through-leakage" does not directly affect the process in the working chamber, but it passes through it from the discharge plenum towards the suction port.

The leaking gas velocity is derived from the momentum equation, which accounts for the fluid-wall friction:

$$w_l dw_l + \frac{dp}{\rho} + f\frac{w_l^2}{2}\frac{dx}{D_g} = 0 \quad (3.12)$$

where $f(\mathrm{Re}, \mathrm{Ma})$ is the friction coefficient which is dependent on the Reynolds and Mach numbers, D_g is the effective diameter of the clearance gap, $D_g \approx 2\delta_g$ and dx is the length increment. From the continuity equation and assuming that $T \approx$ const to eliminate gas density in terms of pressure, the equation can be integrated in terms of pressure from the high pressure side at position 2 to the low pressure side at position 1 of the gap to yield:

$$\dot{m}_l = \rho_l w_l A_g = \sqrt{\frac{p_2^2 - p_1^2}{a^2\left(\zeta + 2\ln\frac{p_2}{p_1}\right)}} \quad (3.13)$$

where $\zeta = fL_g/D_g + \Sigma\xi$ characterizes the leakage flow resistance, with L_g clearance length in the leaking flow direction, f friction factor and ξ local resistance coefficient. ζ can be evaluated for each clearance gap as a function of its dimensions and shape and flow characteristics. a is the speed of sound.

The full procedure requires the model to include the friction and drag coefficients in terms of Reynolds and Mach numbers for each type of clearance.

Likewise, the working fluid friction losses can also be defined in terms of the local friction factor and fluid velocity related to the tip speed, density, and elementary friction area. At present the model employs the value of ζ in terms of a simple function for each particular compressor type and use. It is determined as an input parameter.

These equations are incorporated into the model of the compressor and employed to compute the leakage flow rate for each clearance gap at the local rotation angle θ.

3.1.4 Oil or Liquid Injection

Injection of oil or other liquids for lubrication, cooling or sealing purposes, modifies the thermodynamic process in a screw compressor substantially. The following paragraph outlines a procedure for accounting for the effects of oil injection. The same procedure can be applied to treat the injection of any other liquid. Special effects, such as gas or its condensate mixing and dissolving in the injected fluid or vice versa should be accounted for separately if they are expected to affect the process. A procedure for incorporating these phenomena into the model will be outlined later.

A convenient parameter to define the injected oil mass flow is the oil-to-gas mass ratio, $m_{\mathrm{oil}}/m_{\mathrm{gas}}$, from which the oil inflow through the open oil port, which is assumed to be uniformly distributed, can be evaluated as

$$\dot{m}_{\mathrm{oil}} = \frac{\dot{m}_{\mathrm{oil}}}{\dot{m}_{\mathrm{gas}}}\dot{m}\frac{z_1}{2\pi} \quad (3.14)$$

where the oil-to-gas mass ratio is specified in advance as an input parameter.

In addition to lubrication, the major purpose for injecting oil into a compressor is to cool the gas. To enhance the cooling efficiency the oil is atomized into a spray of fine droplets by means of which the contact surface between the gas and the oil is increased. The atomization is performed by using specially designed nozzles or by simple high-pressure injection. The distribution of droplet sizes can be defined in terms of oil-gas mass flow and velocity ratio for a given oil-injection system. Further, the destination of each distinct size of oil droplets can be followed until it hits the rotor or casing wall by solving the dynamic equation for each droplet size in a Lagrangian frame, accounting for inertia gravity, drag, and other forces. The solution of the droplet energy equation in parallel with the momentum equation should yield the amount of heat exchange with the surrounding gas.

In the present model, a simpler procedure is adopted in which the heat exchange with the gas is determined from the differential equation for the instantaneous heat transfer between the surrounding gas and an oil droplet. Assuming that the droplets retain a spherical form, with a prescribed Sauter mean droplet diameter d_S, the heat exchange between the droplet and the gas can be expressed in terms of a simple cooling law $Q_o = h_o A_o (T_{\text{gas}} - T_{\text{oil}})$, where A_o is the droplet surface, $A_o = d_S^2 \pi$, d_S is the Sauter mean diameter of the droplet and h_o is the heat transfer coefficient on the droplet surface, determined from an empirical expression. The exchanged heat must balance the rate of change of heat taken or given away by the droplet per unit time, $Q_o = m_o c_{\text{oil}} dT_o/dt = m_o c_{\text{oil}} \omega dT_o/d\theta$, where c_{oil} is the oil specific heat and the subscript o denotes oil droplet. The rate of change of oil droplet temperature can now be expressed as:

$$\frac{dT_o}{d\theta} = \frac{h_o A_o \left(T_{\text{gas}} - T_o\right)}{\omega m_o c_{\text{oil}}} \tag{3.15}$$

The heat transfer coefficient h_o is obtained from:

$$Nu = 2 + 0.6\,\text{Re}^{0.6}\text{Pr}^{0.33} \tag{3.16}$$

Integration of the equation in two time/angle steps yields the new oil droplet temperature at each new time/angle step:

$$T_o = \frac{T_{\text{gas}} - kT_{o,p}}{1+k} \tag{3.17}$$

where $T_{o,p}$ is the oil droplet temperature at the previous time step and k is the non-dimensional time constant of the droplet, $k = \tau/\Delta t = \omega\tau/\Delta\theta$, with $\tau = m_o c_{\text{oil}}/h_o A_o$ being the real time constant of the droplet. For the given Sauter mean diameter, d_S, the non-dimensional time constant takes the form

$$k = \frac{\omega m_o c_{\text{oil}}}{h_o A_o \Delta\theta} = \frac{\omega d_S c_{\text{oil}}}{6 h_o \Delta\theta} \tag{3.18}$$

The derived droplet temperature is further assumed to represent the average temperature of the oil, i.e. $T_{\text{oil}} \approx T_o$, which is further used to compute the enthalpy of the gas-oil mixture.

The above approach is based on the assumption that the oil-droplet time constant τ is smaller than the droplet travelling time through the gas before it hits the rotor or casing wall, or reaches the compressor discharge port. This means that heat exchange is completed within the droplet travelling time through the gas during compression. This prerequisite is fulfilled by atomization of the injected oil. This produces sufficiently small droplet sizes to gives a small droplet time constant by choosing an adequate nozzle angle, and, to some extent, the initial oil spray velocity. The droplet trajectory computed independently on the basis of the solution of droplet momentum equation for different droplet mean diameters and initial velocities. Indications are that for most screw compressors currently in use, except, perhaps for the smallest ones, with typical tip speeds of between 20 and 50 m/s, this condition is well satisfied for oil droplets with diameters below 50 μm. For more details refer to Stosic et al., 1992.

Because the inclusion of a complete model of droplet dynamics would complicate the computer code and the outcome would always be dependant on the design and angle of the oil injection nozzle, the present computation code uses the above described simplified approach. This was found to be fully satisfactory for a range of different compressors. The input parameter is only the mean Sauter diameter of the oil droplets, d_S and the oil properties – density, viscosity and specific heat.

3.1.5 Computation of Fluid Properties

In an ideal gas, the internal thermal energy of the gas-oil mixture is given by:

$$U = (mu)_{\mathrm{gas}} + (mu)_{\mathrm{oil}} = \frac{mRT_{\mathrm{gas}}}{\gamma - 1} + (mcT)_{\mathrm{oil}} = \frac{pV}{\gamma - 1} + (mcT)_{\mathrm{oil}} \tag{3.19}$$

where R is the gas constant and γ is adiabatic exponent.

Hence, the pressure or temperature of the fluid in the compressor working chamber can be explicitly calculated by input of the equation for the oil temperature T_{oil}:

$$T = (\gamma - 1)\frac{(1+k)\,U - (mcT)_{\mathrm{oil}}}{(1+k)\,mR + (mc)_{\mathrm{oil}}} \tag{3.20}$$

If k tends 0, i.e. for high heat transfer coefficients or small oil droplet size, the oil temperature fast approaches the gas temperature.

In the case of a real gas the situation is more complex, because the temperature and pressure can not be calculated explicitly. However, since the internal energy can be expressed as a function of the temperature and specific volume only, the calculation procedure can be simplified by employing the internal energy as a dependent variable instead of enthalpy, as often is the practice. The equation of state $p = f_1(T, V)$ and the equation for specific internal energy $u = f_2(T, V)$ are usually decoupled. Hence, the temperature can be calculated from the known specific internal energy and the specific volume obtained from the solution of differential equations, whereas the pressure can be calculated explicitly from the temperature and the specific volume by means of the equation of state. Typical equations are given in Appendix C.

3.1.6 Solution Procedure for Compressor Thermodynamics

To summarize, the description of the thermodynamic processes in a screw machine is completed by the differential equations for the lobe volumes, which define $V(\theta)$ and $dV/d\theta$, by the differential equation of the internal thermal energy, and by the differential equations describing the working chamber mass balance. In addition the algebraic equations of state and specific internal energy and specific enthalpy, are sufficient to obtain the mass flows through the suction and discharge ports and through the clearances, the mass in the working chamber, the pressure and temperature of the fluid in the working chamber and the mass and the temperature of the injected oil.

If the fluid states described by the pressure and temperature in the pressure and suction plenums are considered to vary with the rotation angle, it is necessary to couple the differential equations for the energy and mass flow rates. The total number of differential equations is increased now by another two for each plenum.

All the differential equations are solved by means of the Runge-Kutta fourth order procedure. As the initial conditions are arbitrarily selected, the convergence of the solution is achieved after the difference between the two consecutive compressor cycles show a sufficiently small monitoring value prescribed in advance.

The instantaneous bulk density ρ is obtained from the instantaneous mass trapped in the control volume and the size of the corresponding instantaneous volume V as $\rho = m/V$.

The equations of energy and continuity are solved to obtain $U(\theta)$ and $m(\theta)$. Together with $V(\theta)$, the specific internal energy and specific volume $u = U/m$ and $v = V/m$ are now known. T and p, or x can be then calculated.

For an ideal gas:

$$T = (\gamma - 1)\,\frac{u}{R} \qquad p = \frac{RT}{v} \tag{3.21}$$

In this case T and p are calculated explicitly.

For a real gas, i.e. refrigerant or other:

$$p = f_1(T, V)u = f_2(T, V) \tag{3.22}$$

These equations are usually uncoupled, with T obtained by numerical solution of the equation set, where p is obtained explicitly from the equation of state.

For a wet vapour:

In the case of a phase change during a compression or expansion process, the specific internal energy and volume of the liquid-gas mixture are:

$$u = (1 - x)\,u_f + xu_g \qquad v = (1 - x)\,v_f + xv_g \tag{3.23}$$

where u_f, u_g, v_f and v_g are the specific internal energy and volume of liquid and gas and they are functions of saturation temperature only. The equations require an implicit numerical procedure which is usually incorporated in property packages. As a result, temperature T and dryness fraction x are obtained. These equations are of the same form for any kind of fluid, and they are essentially simpler than any others in derived form. Moreover, the inclusion of any other phenomena into the differential equations of internal energy and continuity is straightforward.

3.2 Compressor Integral Parameters

Numerical solution of the mathematical model of the physical processes in a compressor provides a basis for more accurate computation of all desired integral (bulk) characteristics than the more traditional integral approach. The most important of these properties are the compressor mass flow rate $\dot{m}$[kg/s], the indicated power $\mathrm{P_{ind}}$[kW], specific indicated power $\mathrm{P_s}$[kJ/kg], volumetric efficiency η_v, adiabatic efficiency η_a, isothermal efficiency η_t, other efficiencies, and the power utilization coefficient–indicated efficiency η_i.

The instant fluid mass trapped in the working volume is determined as the difference between the total fluid mass inflow and outflow:

$$m = m_{\mathrm{in}} - m_{\mathrm{out}} \tag{3.24}$$

where m_{in} and m_{out} are obtained from integration of the corresponding differential equations over the cycle. During the rotation of the compressor shaft, due to the differing shaft speeds of the main and gate rotors being compensated by the different number of lobes in each rotor, the total mass within the compressor is mz_1, where z_1 is the number of lobes on the main rotor. Hence the actual fluid mass flow $\dot{m}$[kg/s] is

$$\dot{m} = mz_1n/60 \tag{3.25}$$

where n is the number of revolution per minute of the main rotor.

The volume delivery $\dot{V}$ is defined with reference to the suction conditions and is usually expressed in [m^3/min]:

$$\dot{V} = 60\ \mathrm{m}/\rho_0 \tag{3.26}$$

From the known maximum volume of the working chamber, the theoretical mass flow is:

$$\dot{m}_t = \frac{(F_{1n} + F_{2n})\,Lnz_1\rho}{60} \tag{3.27}$$

where F_{1n} and F_{2n} are the cross-section areas of the lobes in the front plane of the main screw and the gate screw, respectively, and L is the length of the screw. The volumetric efficiency is readily obtained as:

$$\eta_v = \frac{\dot{m}}{\dot{m}_t} \tag{3.28}$$

It is worth noting that the effects of leakage, gas heat exchange and gas retention in the pockets on the pressure side of asymmetric lobe profiles, are all included in the volumetric efficiency coefficient as a result of the differential form of the governing equations of the mathematical model.

The indicated work transferred to the screw rotors during the suction, expansion and discharge processes is represented by the area of the indicated p-V diagram.

$$W_{\text{ind}} = \int_{\text{cycle}} V dp \tag{3.29}$$

Within the indicated work, flow loses during suction, expansion and discharge, leakages and heat exchange, as well as the influence of injected oil have been included in the differential equations of the model in the same way.

The indicated work in a single compressor working chambers is further used for the computation of the compressor indicated power:

$$P_{\text{ind}} = \frac{W_{\text{ind}} z_1 n}{60} \tag{3.30}$$

In addition to the indicated power, it is useful to know the specific indicated power:

$$W_{\text{sind}} = \int_{\text{cycle}} \frac{V}{m} dp \tag{3.31}$$

where m is the mass of the fluid contained in the working chamber V in the considered instant of time.

The indicated work can be compared with the theoretical adiabatic or isothermal work to yield the corresponding efficiency:

$$\eta_t = \frac{W_t}{W_{\text{sind}}} \qquad \eta_a = \frac{W_a}{W_{\text{sind}}} \tag{3.32}$$

Here the theoretical isothermal and adiabatic works are determined from the common theoretical expressions. For an ideal gas, the theoretical isothermal and adiabatic works are respectively:

$$W_t = RT_1 \ln \frac{p_2}{p_1} \qquad W_a = \frac{\gamma}{\gamma - 1} R \left(T_2 - T_1\right) \tag{3.33}$$

where 1 denotes the beginning, and 2 the end conditions of the compression process.

Specific indicated power is obtained from the known indicated power and delivery:

$$P_{\sin d} = \frac{P}{\dot{V}} \tag{3.34}$$

3.3 Pressure Forces Acting on Screw Compressor Rotors

Screw compressor rotors are subjected to severe pressure loads. The rotors, as well as their bearings must satisfy rigidity and elasticity requirements to ensure reliable compressor operation.

In order to explain the calculation of the pressure loads, several cases are considered. Let the pressure $p(\theta)$ be known for any instantaneous angle of rotation θ, with a reasonable angle increment, say 1 degree. Figure 3.1 presents the radial and torque forces in a rotor cross section. The example is given for 5/6 "N" rotors. The pressure p acts in the corresponding interlobes normal to line AB. A and B are either on the sealing line between rotors or on the rotor tips. Since they belong to the sealing line, they are fully defined from the rotor geometry.

In position 1, there is no contact between the rotors. Since A and B are on a circle, the overall forces F_1 and F_2 act towards the rotor axes and they are purely radial. There is no torque caused by pressure forces in this position.

In position 2, there is only one contact between the rotors at point A. Forces F_1 and F_2 are eccentric and comprise two both radial and circumferential components. The latter cause the torque. Due to the force position, the torque on the gate rotor is significantly smaller then that on the main rotor.

In position 3, both contact points are on the rotors with equal overall and radial forces for both rotors. As in the previous case, they also cause torque.

3.3.1 Calculation of Pressure Radial Forces and Torque

Let the x direction be parallel to the line between rotor axes O_1 and O_2 with y is perpendicular to x. The radial force components are:

$$
\begin{aligned}
R_x &= -p \int_A^B dy = -p\,(y_B - y_A) \\
R_y &= -p \int_A^B dx = -p\,(x_B - x_A)
\end{aligned}
\tag{3.35}
$$

The torque is:

$$
T = p \int_A^B x dx + p \int_A^B y dy = 0.5p\left(x_B^2 - x_A^2 + y_B^2 - y_A^2\right) \tag{3.36}
$$

The above equations are integrated along the profile for all profile points. Then they are integrated for all angle steps to complete one revolution employing a given pressure history $p = p(\theta)$. Finally, the sum for all rotor interlobes is

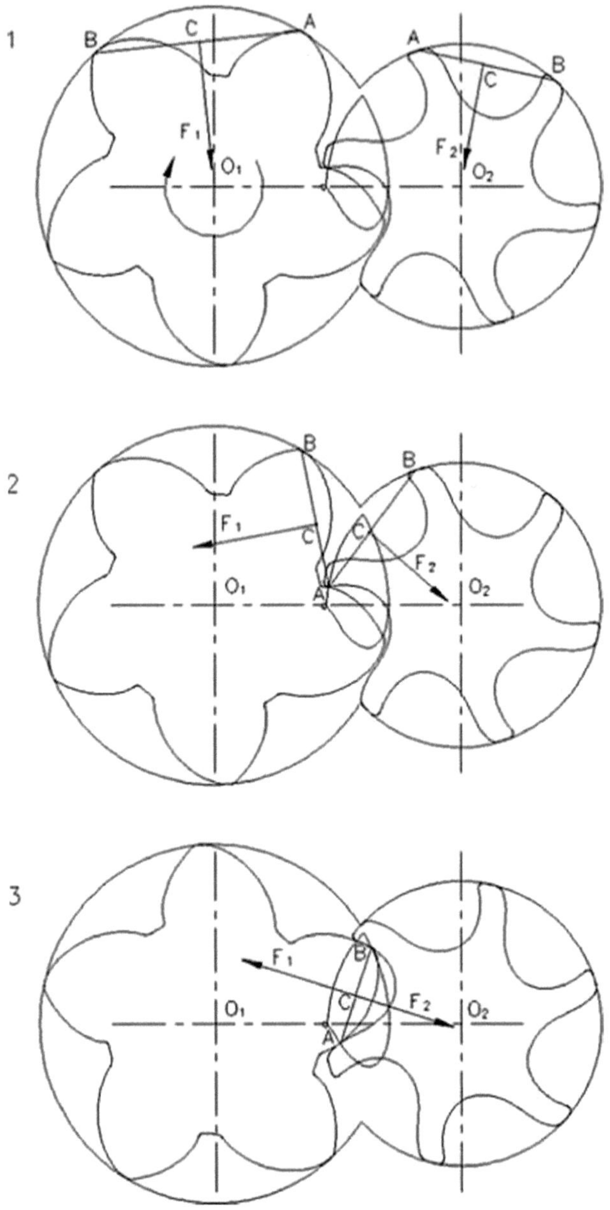

Fig. 3.1. Pressure forces acting on screw compressor rotors

made taking into account the phase shift, as well as the axial shift between the interlobes.

Since the gate rotor has a larger lead angle than the main one, proportional to the gearing ratio z_2/z_1, where z is the number of rotor lobes, appropriate summation usually leads to larger radial forces on the gate rotor despite the fact that the gate rotor may be smaller then the main one, as shown in Fig 3.2.

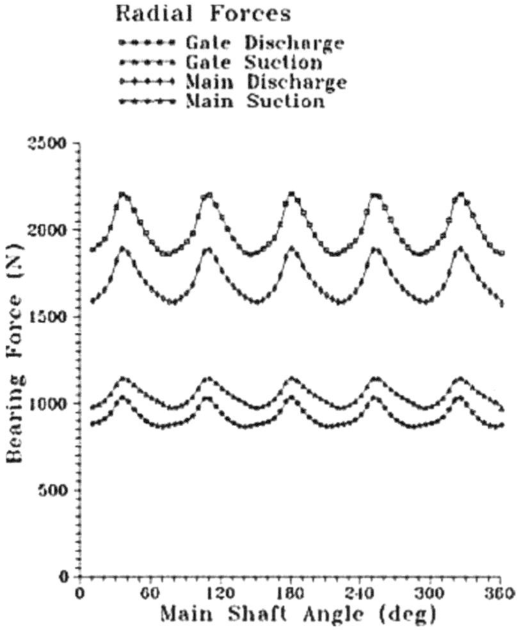

Fig. 3.2. Bearing radial forces

The axial force, shown in Fig. 3.3 is the product of the pressure and interlobe cross section. In some regions the interlobes overlap each other. The gate rotor covers a small part of the main rotor interlobe, while the main rotor covers the majority of the gate interlobe. This phenomenon causes the main axial force to be disproportionally larger than the gate one. A correction is allowed for axial forces which takes into account the fact that the pressure in the rotor front gaps also acts in the axial direction by using an average of pressures in two neighbouring interlobes to act on the lobe in question.

Rotor axial forces, being offset from their axes of rotation, act to minimize the discharge bearing radial force and increase the suction bearing forces. This is generally convenient, because the suction bearing forces are usually smaller than the discharge ones. Bearing in mind that the main rotor axial force is larger than that on the gate, this effect is more beneficial for the main rotor.

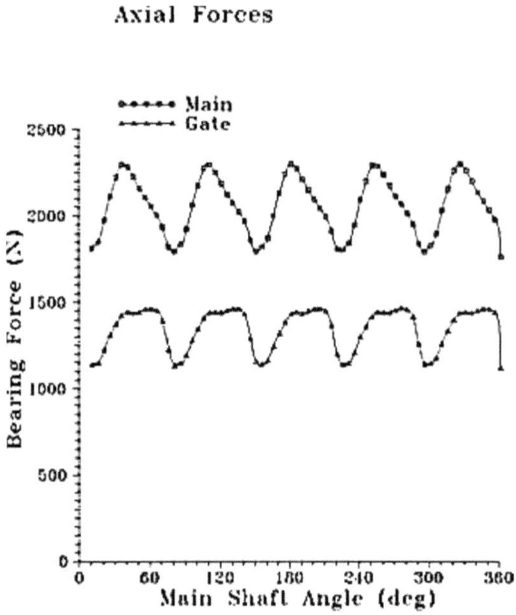

Fig. 3.3. Bearing axial forces

Bearing radial and axial forces and rotor torque have been calculated for a 5/6-128 mm oil-flooded air screw compressor rotors for an inlet pressure of 1 bar and a discharge pressure of 8 bar.

The results of the estimates of the radial reactions of the suction and discharge bearings, torque, axial forces, angular and axial positions of the radial force as defined in Fig. 3.4, are all given as a function of the rotation angle.

3.3.2 Rotor Bending Deflections

Bearing radial reactions are calculated by means of the procedure presented in Fig. 3.5. This represents rotors subjected to a radial force R with bearing reactions R_D and R_S on the discharge and suction rotor ends over a span z_2. A rotor elastic line function is given by differential equation:

$$\frac{d^2\delta}{dz^2} = \frac{M}{EI}$$

where $\delta = \delta(z)$ is the deflection, M is a bending moment function, E is the modulus of elasticity $E = 2.1\ 10^{11} Pa$ and I is a rotor polar momet of inertia, calculated from the rotor geometry, by means of numerical integration. Since the radial force R and its axial position z_1 are calculated in advance for every rotation angle, d is also a function of the rotation angle.

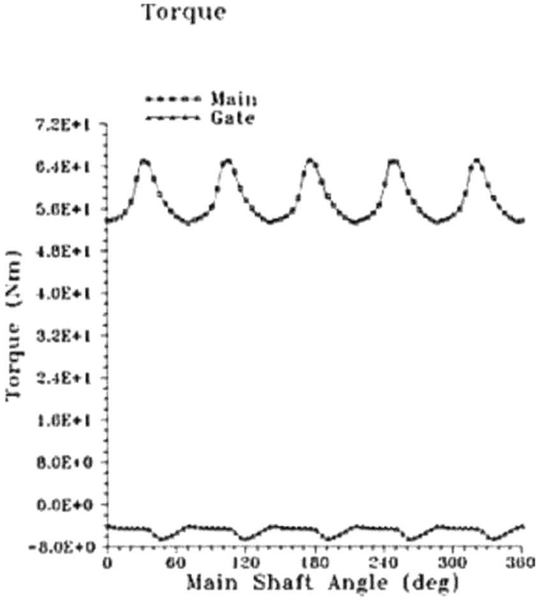

Fig. 3.4. Rotor torque

Integration of this equation over the rotor span between the two radial bearings gives the bending deflections as functions of the rotor axial coordinate z which has its own maximum value. This is calculated for every increment of the rotation angle.

3.4 Optimisation of the Screw Compressor Rotor Profile, Compressor Design and Operating Parameters

The mathematical modelling and computer simulation techniques, as described in Sects. 3.1 and 3.2 enable rapid assessments to be made of the effects of changes in the rotor profile or proportions on compressor performance. Other investigators such as Tang and Fleming, 1992, 1994, Sauls, 1994, Fleming, 1994 and Fujiwara and Osada, 1995 have given examples of how rotor design can be optimised by this means. The authors have, however, taken this a stage further to develop more generalised optimisation procedures to improve the entire screw compressor design, as first shown by Hanjalic and Stosic, 1997.

3.4.1 Optimisation Rationale

A problem in optimisation is the number of calculations which must be performed to identify and reach the desired value. Another problem is how to

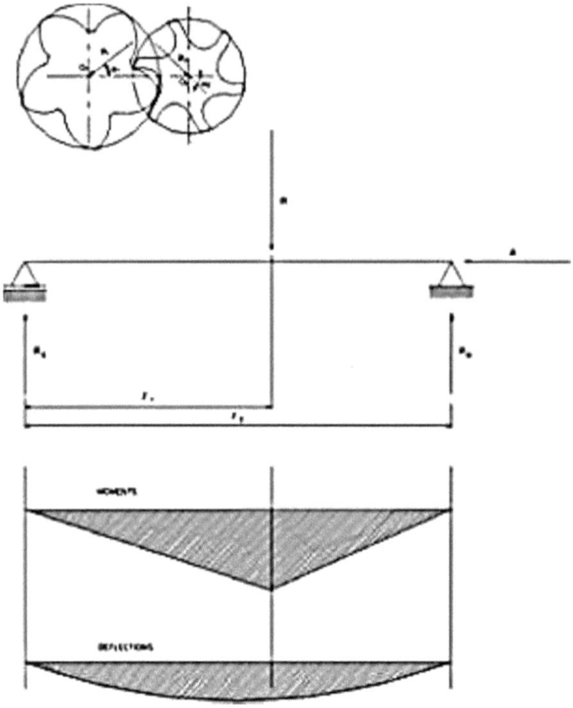

Fig. 3.5. Bending forces, moments and deflections

be certain that the optimum calculated is the global optimum. Among the optimisation methods frequently used in engineering are those of steepest descent, Newton's method, Davidon-Fletcher-Powell's method, random search, grid search method, search along coordinate axes, Powell's method and Hooke-Jeeves's method. A widely used method for the optimisation of functions with several optima is the genetic algorithm. It requires only a value of the target function and it can conveniently handle discontinuities, however this method is slow in converging to a solution. Alternatively a constrained simplex method, known as the Box complex method can be conveniently used. It also requires the function value only and not its gradient. The disadvantage is that it is less suitable for discrete parameters, for example, if a choice between discrete component sizes is required.

The box complex method was therefore used here to find the local minima, which were input to an expanding compressor database. This finally served to estimate a global minimum. That database may be used later in conjunction with other results to accelerate the minimization.

The constrained simplex method emerged from the evolutionary operation method which was introduced already in the 1950s by Box, 1957 and Box and Draper, 1969. The basic idea is to replace the static operation of a process by a continuous and systematic scheme of slight perturbations in the control variables. The effect of these perturbations is evaluated and the process is shifted in the direction of improvement. The basic simplex method was originally developed for evolutionary operation, but it was also suitable for the constrained simplex method. Its main advantage is that only a few starting trials are needed, and the simplex immediately moves away from unsuitable trial conditions. The simplex method is especially appropriate when more than three control variables are to be perturbed and the process requires a fresh optimisation with each new set of input data.

There are several criteria for screw profile optimisation which are valid irrespective of the machine type and duty. Thus, an efficient screw machine must admit the highest possible fluid flow rates for a given machine rotor size and speed. This implies that the fluid flow cross-sectional area must be as large as possible. In addition, the maximum delivery per unit size or weight of the machine must be accompanied by minimum power utilization for a compressor and maximum power output for an expander. This implies that the efficiency of the energy interchange between the fluid and the machine is a maximum. Accordingly unavoidable losses such as fluid leakage and energy losses must be kept to a minimum. However, under some conditions increased leakage may be more than compensated by greater bulk fluid flow rates. Specification of the required compressor delivery rate requires simultaneous optimisation of the rotor size and speed to minimise the compressor weight while maximising its efficiency. Finally, for oil-flooded compressors the oil injection flow rate, inlet temperature and position needs to be optimised. It follows that a multivariable minimisation procedure is needed for screw compressor design with the optimum function criterion comprising a weighted balance between compressor size and efficiency or specific power.

3.4.2 Minimisation Method Used in Screw Compressor Optimisation

The power and capacity of contemporary computers is only just sufficient to enable a full multivariable optimisation of both the rotor profile and the whole compressor design to be performed simultaneously in one pass.

The optimisation of a screw compressor design is generically described as a multivariable constrained optimisation problem. The task is to maximise a target function $f(x_1, x_2, \ldots, x_n)$, subjected simultaneously to the effects of explicit and implicit constraints and limits, $g_i \leq x_i \leq h_i$, $i = 1, n$ and $g_i \leq y_i \leq h_i$, $i = n+1, m$ respectively, where the implicit variables $y_{n+1}, \ldots, y_m$ are dependent functions of x_i. The constraints g_i and h_i are either constants or functions of the variables x_i.

When attempting to optimise a compressor design a criterion for a favourable result must be decided, for instance the minimum power consumption, or operating cost. However, the power consumption is coupled to other requirements which should be satisfied, for example a low compressor price, or investment cost. The problem becomes obvious if the requirement for low power consumption conflicts with the requirement of low compressor price. For a designer, the balance is often completed with sound judgement. For an optimisation program the balance must be expressed in numerical values. This is normally done by weighting the different parts of the target function.

An example of the usage of weights is the target function $F = w_1 L + w_2 C$, where L is the calculated power loss and C is a measure of the compressor price. The choice of weights may substantially change the target function, and some choices can lead to a target function which is difficult or impossible to optimise. Moreover, it is likely that many combinations of weights w_1 and w_2 will result in a target function with several equally good optima. It is obvious that with a large number of conflicting performance criteria, the tasks of the optimisation program and its user will be more difficult. When using multi-target optimisation, the separate parts of the objective function are evaluated to eliminate some of the difficulties in the definition of the target function.

Another important issue for real-world optimisation problems is constraints. In the general case, there are two types of constraints, explicit or implicit. The explicit ones are limitations in the range of optimised parameters, for example available component sizes. These two different constraints can, in theory, be handled more or less in the same way. In practice, however, they are handled differently.

The implicit constraints are often more difficult to manage than explicit constraints. The most convenient and most common way is to use penalty functions and thus incorporate the constraints in the objective function. Another way is to tell the optimisation algorithm when the evaluated point is invalid and generate a new point according to some predetermined rule.

Generally, it can be said that constraints, especially implicit constraints, make the optimisation problem harder to solve, since it reduces the solution space.

In the early 1960s, a method called the simplex method emerged as an empirical method for optimisation, this should not be confused with the simplex method for linear programming. The simplex method was later extended by Box, 1965 to handle constrained problems. This constrained simplex method was appropriately called the complex method, from constrained simplex. Since then, several versions have been used. Here, the basic working idea is outlined for the complex method used. If a nonlinear problem is to be solved, it is necessary to use k points in a simplex, where $k = 2n$. These starting points are randomly generated so that both the implicit and explicit conditions are satisfied. Let the points x^h and x^g be defined by

$$\begin{aligned} f(x^h) &= \max f(x^1), f(x^2), \dots, f(x^k) \\ f(x^g) &= \min f(x^1), f(x^2), \dots, f(x^k) \end{aligned} \tag{3.37}$$

calculate the centroid $\bar{x}$ of those points other than x^l by

$$\bar{x} = \frac{1}{k-1} \sum_{i=1}^{k} x_j^i, \quad x^i \neq x^l \tag{3.38}$$

The main idea of the algorithm is to replace the worst point x^l by a new and better point. The new point x^r is calculated as a reflection of the worst point through the centroid. This is done as

$$x^r = \bar{x} + \alpha(\bar{x} - x^l)$$

where the reflection coefficient α is chosen according to Box as $\alpha = 1.3$.

The point x^r is examined with regard to explicit and implicit constraints and if it is feasible x^l is replaced with x^r unless $f(x^r) \leq f(x^l)$. In that case, it is moved halfway towards the centroid of the remaining points. This is repeated until it stops repeating as the lowest value. However, this cannot handle the situation where there is a local minimum located at the centroid. The method used here is to gradually move the point towards the maximum value if it continues to be the lowest value. This will, however, mean that two points can come very close to each other compared to other points, with a risk of collapsing the complex. Therefore, a random value is also added to the new point. In this way, the algorithm will take some extra effort to search for a point with a better value, but in the neighbourhood of the point of the maximum value. It is consequently guaranteed that a point better than the worst of the remaining points will be found. Expressed as an equation

$$x^{r(\text{new})} = 0.5\left[x^{r(\text{old})} + c\bar{x} + (1-c)x^h\right] + (\bar{x} - x^h)(1-c)(2R-1) \qquad (3.39)$$

where

$$c = \left(\frac{n_r}{n_r + k_r - 1}\right)^{\frac{n_r + k_r - 1}{n_r}} \qquad (3.40)$$

k_r is the number of times the point has repeated itself as lowest value and n_r is a constant. Here $n_r = 4$ has been used. R is a random number in the interval [0,1].

If a point violates the implicit constraints, it is moved halfway towards the centroid. In order to handle the case of the centroid violating the implicit constraint, the point is gradually moved towards the maximum value. If the maximum value is located very close to the implicit constraint, this will take many iterations and the new point will be located very close to the maximum value and will not really represent any new information. Therefore a random value is added also in this case. Now

$$c = \left(\frac{n_r}{n_r + k_c - 1}\right)^{\frac{n_r + k_c - 1}{n_r}} \qquad (3.41)$$

where k_c is now the number of times the point has violated the constraint.

These modifications of the complex method have led to a robust method which has already been used in many engineering applications.

The power and capacity of contemporary computers just enable a full multivariable optimisation of both the rotor profile and compressor design in one pass.

The geometry of screw machines is dependent on a number of parameters whose best values to meet specified criteria can, in principle, be determined by a general multi-variable optimization procedure. In practice it is preferable to restrict the number of parameters to a few, which are known to be the most significant, and restrict the optimization to them only.

In the example presented, the rotor centre distance, as well as the main rotor diameter are kept constant, while other profile radii r_0, r_1, r_2 and r_3 are made variable.

An example is given for an air compressor since due to the higher ratio of specific heats, it is more sensitive to design changes than a refrigeration compressor. Two compressor modes were considered, namely a dry air compressor, where the isentropic exponent γ is close to 1.4 and an oil-flooded compressor. A small, 102 mm diameter compressor with axes 72 mm apart was selected to serve as a basis for the optimization. The compressor housing was represented by the built-in volume ratio, while the compressor working parameters were the compressor speed and flow of the injected oil, oil injection position and oil temperature. The air inlet conditions selected were, $p_0 = 1$ bar and $t_0 = 27°\text{C}$, with discharge pressures of 3.5 and 8 bars for the dry and oil-flooded compressor respectively to comply with normal compressor practice. As a result, two distinctively different rotor profiles were calculated, one for oil-free compression, and the other for oil-flooded compression. These are presented in Fig. 3.6.

Although the profiles look alike, there is a substantial difference between their geometry which is given in the following table.

As for any other optimisation, the screw compressor profile and compressor design optimisation must be considered with extreme caution. Namely, since the multivariable optimisation usually finds only local minima, these may not necessarily be globally the best optimisation result. Therefore, an extended calculation should inevitably follow before a final decision on the compressor design is made.

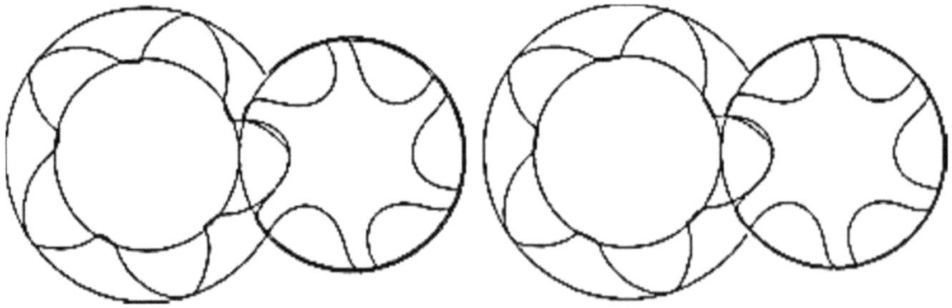

Fig. 3.6. Comparison of two optimized 5-6-102 mm "N" rotors: *left*, dry operation, *right*, oil-flooded operation

	Dry	Oil-Flooded
C [mm]	72	72
r [mm]	18.6	18.6
r_0 [mm]	1.02	0.75
r_1 [mm]	20.5	18
r_2 [mm]	3.5	5.4
r_3 [mm]	6.1	5.8

Further results of the optimisation are:

	Dry	Oil-Flooded
Built-in volume ratio	2,2	4.1
Rotor speed [rpm]	7560	4020
Oil flow [lit/min]	–	12
Injection position [°]	–	66
Oil temperature [°]	–	36

3.5 Three Dimensional CFD and Structure Analysis of a Screw Compressor

It has already been shown that flow through a screw compressor can be described by the mass averaged conservation equations of continuity, momentum, energy and space and an equation of state for the working fluid. Kovacevic, 2002, has shown that the model can be extended to take account of multiphase flow by including a concentration equation and also interactions between the working fluid and the compressor rotors and casing. The numerical solution of such a system of partial differential equations is then made possible by the inclusion of constitutive relations in the form of Stokes' Fick's and Fourier's laws for the fluid momentum, concentration and energy equations, respectively, and Hooke's law for the momentum equations of a thermoelastic body.

These equations are conveniently written in the form of the following generic transport equation:

$$\frac{d}{dt}\int_V \rho\phi dV + \int_S \rho\phi(\mathbf{v}-\mathbf{v}_s)\cdot d\mathbf{s} = \int_S \Gamma_\phi \mathrm{grad}\phi\cdot d\boldsymbol{s} + \int_S \mathbf{q}_{\phi S}\cdot d\boldsymbol{s} + \int_V q_{\phi V}\cdot dV \quad (3.42)$$

Here ϕ stands for the transported variable and Γ_ϕ is the diffusion coefficient. The meaning of the source terms, $\boldsymbol{q}_{\phi S}$ and $q_{\phi V}$ for the fluid and solid transport equations is given in Table 3.1.

The resulting system of partial differential equations is discretised by means of the finite volume method in a general Cartesian coordinate frame. This method enhances the conservation of governing equations while at the same time enables a coupled system of equations to be solved simultaneously for both the solid and fluid regions.

Table 3.1. Terms in the generic transport equation (65)

Equation	ϕ	Γ_ϕ	$\mathbf{q}_{\phi S}$	$Q_{\phi V}$
Fluid Momentum	v_i	μ_{eff}	$\left[\mu_{\text{eff}}(\text{grad}\,\mathbf{v})^{\text{T}} - \left(\frac{2}{3}\mu_{\text{eff}}\text{div}\,\mathbf{v} + p\right)\mathbf{I}\right] \cdot \mathbf{i}_i$	$f_{\text{b},i}$
Solid Momentum	$\frac{\partial u_i}{\partial t}$	η	$\left[\eta(\text{grad}\,\mathbf{u})^{\text{T}} + (\lambda\text{div}\,\mathbf{u} - 3K\alpha\Delta T)\mathbf{I}\right] \cdot \mathbf{i}_i$	$f_{\text{b},i}$
Energy	e	$\frac{k}{\partial e/\partial T} + \frac{\mu_t}{\sigma_T}$	$-\frac{k}{\partial e/\partial T}\frac{\partial e}{\partial p} \cdot \text{grad}p$	$\boldsymbol{T}\colon \text{grad}\,\mathbf{v} + h$
Concentration	c_i	$\rho D_{i,\,\text{eff}}$	0	S_{ci}
Space	$\frac{1}{\rho}$	0	0	0
Turbulent kinetic energy	K	$\mu + \frac{\mu_t}{\sigma_k}$	0	$P - \rho\varepsilon$
Dissipation	ε	$\mu + \frac{\mu_t}{\sigma_\varepsilon}$	$C_1 P\frac{\varepsilon}{k} -$	$C_2\rho\frac{\varepsilon^2}{k} - C_3\rho\varepsilon\text{div}\,\mathbf{v}$

This mathematical scheme is accompanied by boundary conditions for both the solid and fluid parts. The authors introduced a special treatment of the compressor fluid boundaries. The compressor was positioned between two relatively small suction and discharge receivers. By this means, the compressor system is separated from the surroundings by adiabatic walls only. It communicates with its surroundings through the mass and energy sources or sinks placed in these receivers. This way of maintaining constant suction and discharge pressures is given by the authors. The solid part of the system is constrained by both Dirichlet and Neuman boundary conditions through zero displacement in the restraints and zero traction elsewhere. Connection between the solid and fluid parts is determined explicitly through the energy equation if the temperature and displacement from the solid body surface are boundary conditions for the fluid flow and vice versa.

A commercial CCM solver was used to obtain the distribution of pressure, temperature, velocity and the density fields throughout the fluid domain as well as the deformation and stresses of the solid compressor elements. Based on the solution of these equations, integral parameters of screw compressor performance were calculated, as shown by Kovacevic et al. 2003 and Stosic et al., 2003a.

Simultaneous combination of Computational Fluid Dynamics (CFD) and Structure Analysis, Computational Continuum Mechanics (CCM) may be used if the accurate estimation of velocity, pressure, temperature and concentration fields, as well as stress and deformation within a screw compressor is required. To enable this, a rack-generation procedure has been developed to produce the rotor profiles and an analytical transfinite interpolation method to obtain a 3-D numerical mesh. Moreover, adaptive meshing, orthogonalization and smoothing are employed to generate a numerical grid that takes

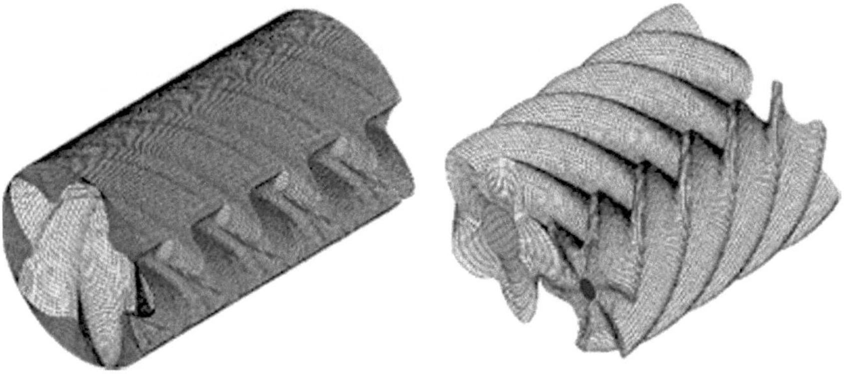

Fig. 3.7. Numerical grid of a screw compressor working chamber – fluid and solid domains

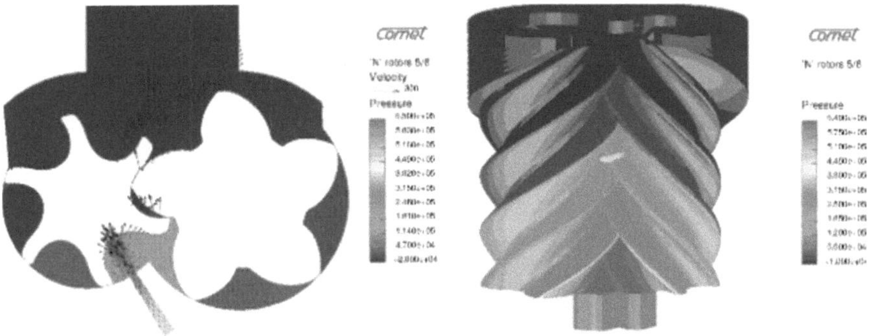

Fig. 3.8. Pressure field on the compressor rotors, and pressure velocity distribution

advantage of the innovative techniques used in recent finite volume numerical method solvers. They were required to overcome problems associated with (i) rotor domains which stretch and slide relative to each other and along the housing (ii) robust calculations in domains with significantly different geometry ranges, (iii) a grid moving technique with a constant number of vertices. These features have been used together to develop an independent standalone CAD-CCM interface program to generate a numerical grid of the screw compressor.

Modifications are also implemented in the calculation procedure to improve solutions in complex domains with strong pressure gradients. Typical results arising from its use when linked to Comet, a commercial CCM solver by ICCM, Germany, are shown for an oil-flooded screw compressor. Examples are given to demonstrate the scope of the method for accurate calculation of processes within these machines.

Some changes within the solver were made to increase the speed of calculation by means of user functions. These include a novel method to maintain

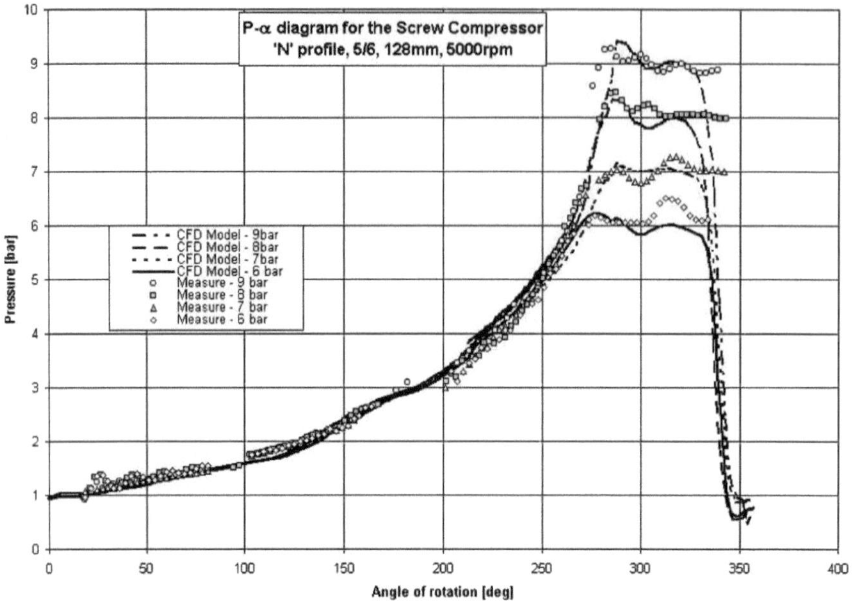

Fig. 3.9. Pressure-angle diagram, comparison with the experimental results

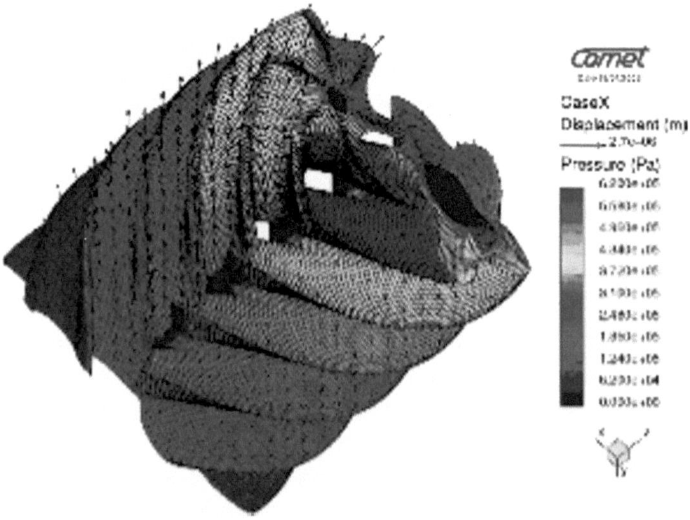

Fig. 3.10. Deformation of screw compressor rotors due to pressure and temperature

constant pressures at the inlet and outlet ports and consideration of two phase flow resulting from oil injection in the working chamber.

The pre-processing code and calculating method have been tested on a commercial CFD solver to obtain flow simulations and integral parameter calculations. Typical results are shown in Fig. 3.7 and Fig. 3.8.

The results of calculations on an oil injected screw compressor are compared with experimental results as shown in Fig. 3.9. Good agreement was obtained between them.

As a result of these developments, full 3-D calculations can now be performed to quantify the interaction of the compressor structure and compressor fluid flow. Distortion diagrams, as shown in Fig. 3.10, can be used to show how working clearances are altered by the effects of temperature and pressure change during the compressor operation.

4

Principles of Screw Compressor Design

As for other design processes, the design of screw compressors is interactive and the measured performance of the compressor is compared with that specified in advance. Usually this is achieved by testing a prototype system and modifying until it is satisfactory. With the help of a simulation model the design process is more reliable, therefore, prototyping can be reduced to a minimum. For that purpose, the authors have developed a suite of subroutines for calculation of the screw compressor performances. This provides facilities for generating new profiles and predicting thermodynamic processes in the compressor and results from it have been verified by extensive comparison with laboratory measurements of the cyclic variation of important local and bulk properties in a screw compressor.

More details of this procedure are given by Stosic et al., 1992. The computational procedure used in this process is divided into the following three phases.

(i) The pre-processor generates the lobe profiles and the complete screw rotor pairs from the algebraically or otherwise prespecified profile curves, or by use of a set of pre-generated points to give a volume function, incorporating all interlobe overlapping and blow hole areas.

(ii) The main program simulates the compressor processes by a set of conservation equations for mass, momentum and energy in one-dimensional differential form which describe the thermodynamic and flow process in an elementary volume within the screw machine at any angle of rotation or instant of time. The integration of the mass and energy equations yields the variation of the pressure and temperature of the gas as a function of the angle of rotation. A pressure-volume relationship is thereby produced, from which the integral compressor characteristics are evaluated.

(iii) The post-processing program supplies the results in graphical form, together with a plot of the rotor pair at any desired angle or sequence of angles of rotation.

4.1 Clearance Management

To perform without causing excessive noise, a compressor must run slowly. Moreover, speed variation is gradually becoming the main method of controlling the capacity of refrigeration compressors. This requires good efficiency at low speeds. This implies that a refrigeration screw compressor must be built with small rotor and housing clearances to maintain low leakage and consequent high efficiency under these conditions. The manufacture of rotors by grinding, especially with simultaneous measurement, control and correction of the profile, makes it possible today to maintain a profile tolerance of 5 μm which, in turn, enables the clearances between the rotors to be kept below 15 μm. With such small clearances, rotor contact is very likely and hence the profile and its clearance distribution must be generated in such a manner that damage or seizure will be avoided should this occur. Therefore the clearance distribution should be calculated with full allowance for the effects of thermal distortion. One such case is presented in Fig. 4.1 left, which represents the clearance distribution on the cold rotors. After the rotor achieves working temperature conditions, their clearance is presented in middle and right. According to the gate rotor torque sign, rotor contact may be obtained either on the round or flat lobe flank middle and right respectively. Round flank contact is traditional and is usually obtained by maintaining a high, so called positive, gate rotor torque. Flat face contact is obtained by maintaining, so called negative, gate rotor torque. It offers advantages which have not yet been widely appreciated. Namely, since the sealing line is longer on the flat

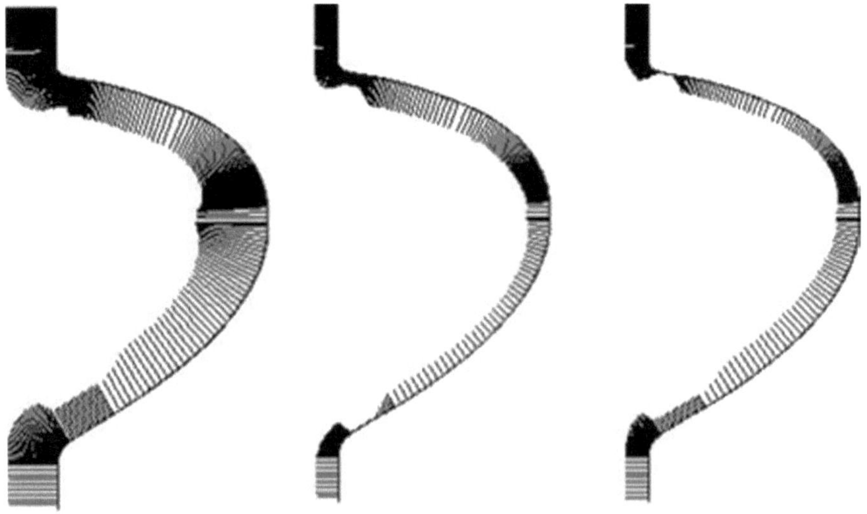

a) Cold rotors b) Rotor contact on the round face c) Contact on the flat face

Fig. 4.1. Interlobe clearance distribution

face, as may be seen, a smaller clearance on that side, which is the case with flat face rotor contact, is very welcome. It decreases interlobe leakage for the same sealing line length and for the same clearance gaps compared with round flank contacts. This consequently decreases the interlobe leakage. Also in the case of negative gate torque, the resulting gate rotor lobe is thicker and the rotor displacement is higher. All these effects lead to higher compressor flows and efficiencies.

There is one additional design aspect, which although important, is not widely appreciated. This is that if bearing clearances are not taken fully into account, small rotor clearances and high pressure loads together can cause contact due to the resulting rotor movement. This contact may occur between the rotor tips and the housing unless the bearing centre distance is smaller than that of the rotor housing. To maintain the rotor interlobe clearance as small as possible, the bearing centre distance must be even further reduced.

Since modern refrigeration compressors rotate slowly or with a reduced suction volume for the majority of their operating time, all their leakages must be minimised. Thus refrigeration compressor rotors are profiled for the smallest possible blow-hole area. However, reduction of the blow-hole area is associated with increase in the sealing line length. It is therefore necessary to find the optimum profile shape which minimises the sum of both the blow-hole and sealing line leakage areas.

4.1.1 Load Sustainability

A general feature of screw compressors is that the pressure difference through them causes high rotor loads and this is especially the case for low temperature refrigeration compressors, where these are large. Therefore, to maintain their rigidity and minimise deflection, the rotors are regularly profiled with a relatively small main rotor addendum in order to increase the gate root diameter. This sometimes leads to very shallow and clumsy rotors. An alternative possibility is to increase the gate rotor lobe thickness. This greatly increases the rotor moment of inertia and thereby reduces the rotor deflection more effectively.

In some compressor designs, multiple cylinder roller bearings or multiple ball bearings are located at the high pressure end of the rotors to withstand the large radial forces reliably over a long operating life. Frequently, two or more bearings are also employed for axial loads. Since only one axial bearing works, the role of the other is usually to clamp the rotor and prevent it bouncing in the axial direction.

In Fig. 4.2 a compressor design is presented with a combination of 11 bearings and a balance piston. It is claimed that this arrangement gives 2–3 times the bearing life of a traditional design.

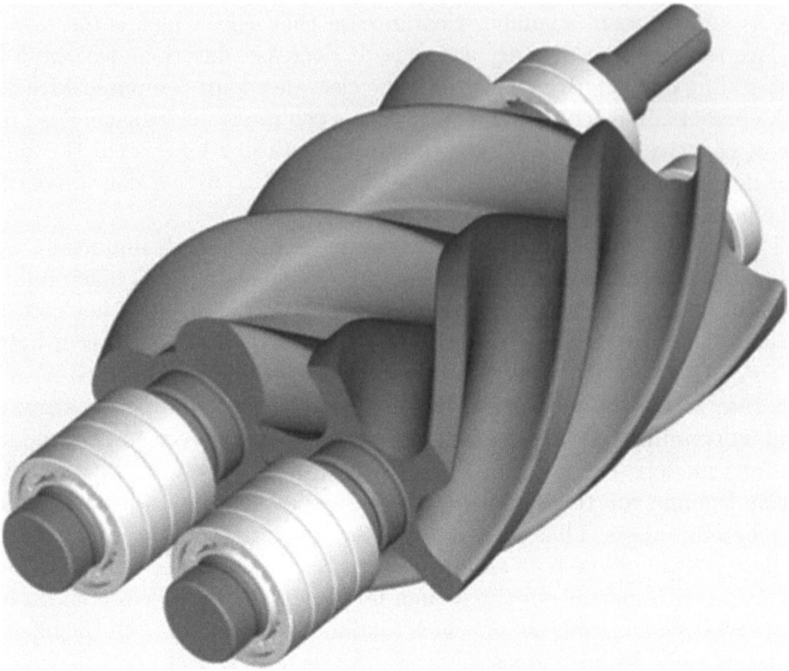

Fig. 4.2. Multibearing arrangement

4.1.2 Compressor Size and Scale

As has already been pointed out, the lower limit for the use of screw compressors is determined by the fact that their efficiencies decrease with size. This is mainly due to the fact that the volumetric throughput is proportional to the cube of their linear dimensions while leakages are proportional to the square. Therefore, compressor leakages and hence their efficiencies are inversely proportional to the compressor linear dimensions. Moreover, since the clearance gaps reach a manufacturing limit, which cannot be scaled down further with the rotor size, leakages become increasingly high in small compressors.

However, since it is likely that manufacturing improvements may lead to smaller compressor clearances in the future, small screw compressors may be introduced which will operate with reasonable efficiencies. An example of what is, possibly, the smallest screw compressor available on the contemporary market, is given in Fig. 4.3. This delivers about 70 lit/min at 3000 rpm and is used at present to compress air. However, in time, this may be extended to refrigeration applications.

The scaling down of screw compressors may lower the limits of their use, which will, bring them into increasing competition with scroll compressors. The outcome of this will affect not only screw, but also scroll compressors.

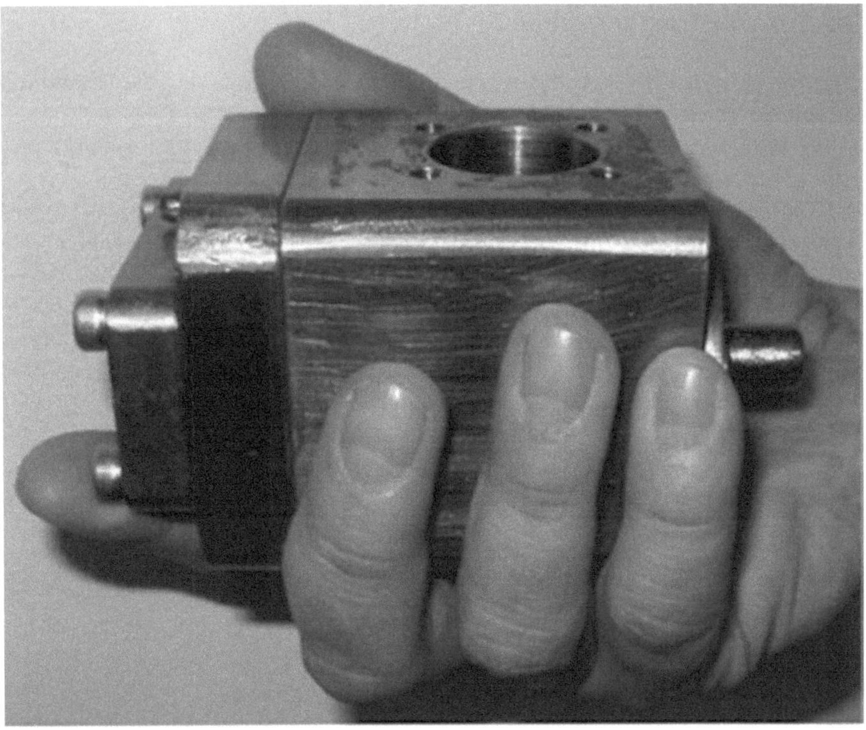

Fig. 4.3. Small screw compressor based on 35 mm 4-5 rotors, displacement $30\,\mathrm{cm}^3/\mathrm{rev}$

4.1.3 Rotor Configuration

It is well known, that increasing the number of lobes enables the same built-in volume ratio to be attained with larger discharge ports. Larger discharge ports decrease the discharge velocity and therefore reduce the discharge pressure losses, thereby increasing the compressor overall efficiency. Hence refrigeration compressors tend to be built with more lobes than the traditional 4-6 combination and 5-6, 5-7 and 6-7 configurations are becoming increasingly popular. Also, the greater the number of lobes, the smaller the pressure difference between the two neighbouring working chambers. Thus, rotor tip leakage and blow-hole losses are reduced. Furthermore, more lobes combined with a large wrap angle ensure multiple rotor contacts which reduce vibrations and thus minimise noise.

However, more lobes usually mean less rotor throughput and longer sealing lines for the same rotor size, which implies that refrigeration compressors are somewhat larger than their air counterparts. Also more lobes increase the cost of manufacture.

4.2 Calculation Example: 5-6-128 mm Oil-Flooded Air Compressor

An oil-flooded air compressor based on "N" rotors was chosen as an example for the calculation of geometrical, thermodynamic and design characteristics. The design specification was for an air delivery of 7 m^3/min at 8 bar (abs). The maximum flow was to be approximately 10 m^3/min and the highest pressure 15 bar (abs). Manufacturing facilities limited the rotor diameter to 127.5 mm. To obtain the most favourable performance for an oil flooded air compressor with this duty, the rotor combination chosen was 5 lobes on the main and 6 lobes on gate rotor. Due to the high maximum pressure required, the maximum length to diameter ratio (L/D) was limited to 1.65:1.

The calculated compressor displacement was 1.56 lit/rev. The sealing line was 0.13 m for one interlobe and the blow-hole area was 1.85 mm^2.

Firstly, the rotor profile was generated according to the principles described in Chap. 3. Calculation of the thermodynamic performance and resulting forces was then carried out with the software design package developed by the authors, which as a suite of subroutines served to estimate flow, power, specific power and compressor volumetric and adiabatic efficiencies. All design shapes and the component sizes thus derived were transferred to a CAD system by means of a full internal interface and these are shown in Figs. 4.4 and 4.5.

The results showed the rotor swept volume to be 1.56 lit/rev and that the design flow rate would be attained at 4800 rpm for an estimated power input of less than 40 kW. The estimated specific power input was therefore only 5.7 kW/m^3/min.

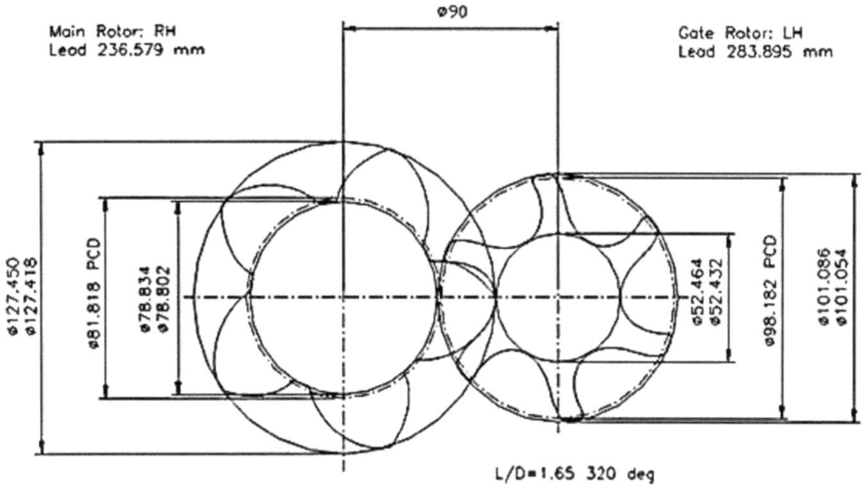

Fig. 4.4. 5-6-128 mm "N" rotors

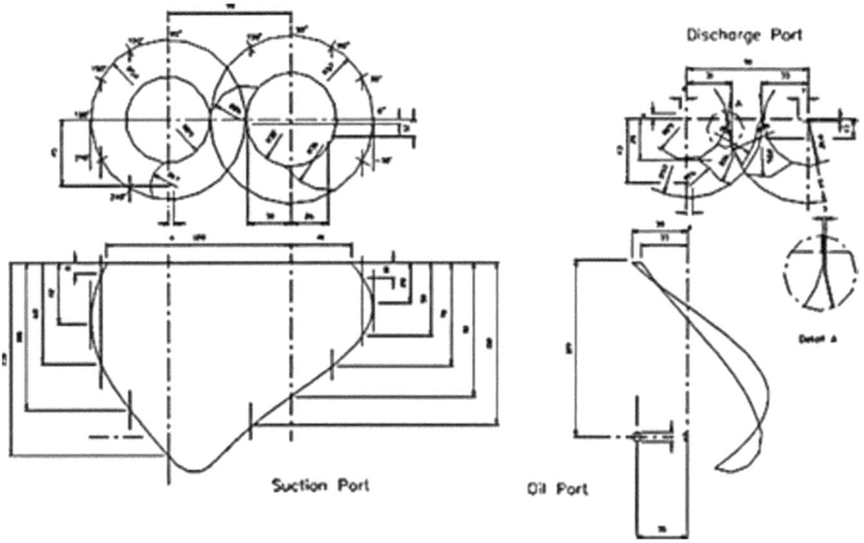

Fig. 4.5. Suction, oil and discharge port of the 5-6-128 mm compressor

Although advanced rotor profiles are a necessary condition for a screw compressor to be efficient, all other components must be designed to enhance the rotor superiority if full advantage of their benefits is to be achieved. Thus rotor to housing clearances, especially at the high pressure end must be properly selected. This in turn requires either expensive bearings with small clearances or cheaper bearings with their clearances reduced to an acceptable value by

Fig. 4.6. 5-6-128 mm oil-flooded air compressor

preloading. The latter practice was chosen as the most convenient and economic solution.

Special care was given to minimise the flow losses in the suction and discharge ports.

The suction port was positioned in the housing to let the air enter with the fewest possible bends and the air approach velocity was kept low by making the flow area as large as possible. The photograph in Fig. 4.6 demonstrates this feature.

The discharge port size was first determined by estimating the built-in-volume ratio required for optimum thermodynamic performance. It was then increased in order to reduce the exit air velocity and hence obtain the minimum combination of internal and discharge flow losses. The port drawing is presented in Fig. 4.5.

The cast iron casing, which was carefully dimensioned to minimize its weight, contained a reinforcing bar across the suction port to improve its rigidity at higher pressures. After casting, it was hydraulically tested at a pressure of 22.5 bar.

A comparison of the calculated and measured data for this compressor is given in the following section.

4.2.1 Experimental Verification of the Model

Before a computer program may be used for performance prediction or design optimisation, the assumed models of all flow effects within it must be verified either independently or within the computer program as a whole. In this case, predictions of all significant bulk and local changes within a screw machine

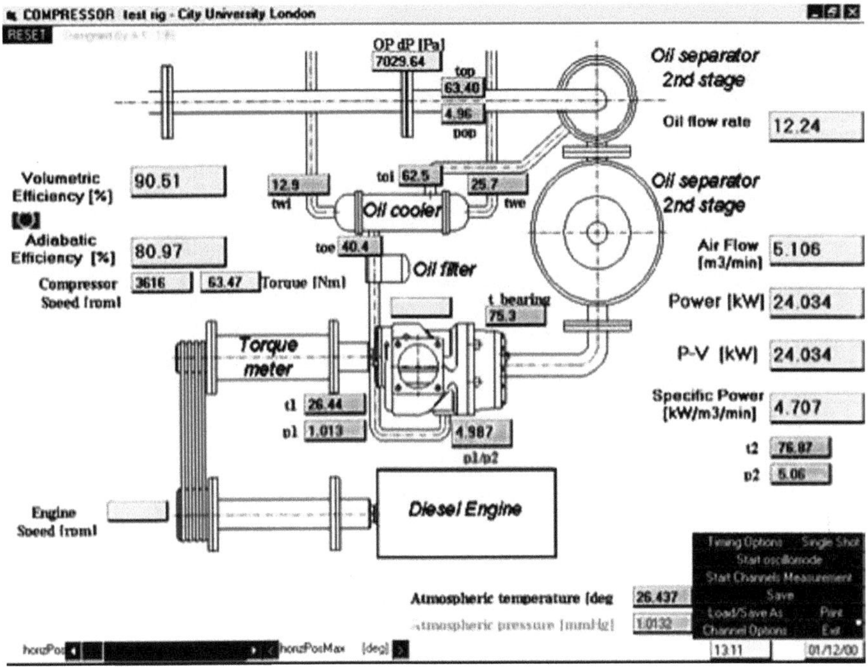

Fig. 4.7. Compressor test layout on the computer screen

were compared with extensive laboratory tests on a screw air compressor. These were carried out over several years in parallel with the development of the mathematical model until satisfactory agreement was obtained between the predicted and measured results. As an insight into the procedure, the method of p-V measurement is presented here. More details can be found in Stosic et al., 1992.

A comparison of the model and experimental results is presented in Fig. 4.9.

An experimental test rig was designed and built at City University, London, to meet Pneurop/Cagi requirements for screw compressor acceptance tests and these were all carried out in accordance with ISO 1706, with delivery flow measurements taken according to BS5600. A 100 kW diesel engine was used as a variable speed prime mover and this enabled oil-flooded compressors of up to 16 m^3/min to be tested. A photograph of the rig is given in Fig. 4.8.

Due to the pulsating character of flow within a screw compressor, the instruments used must have both a fast response rate and a wide frequency range. They must also be small in order to be located in a restricted space without their presence affecting the compressor operation. Accordingly a method was developed to obtain the entire p-V diagram from time averaged measurements of only 4 pressure transducers of the piezzo resistive type,

Fig. 4.8. 5-6-128 mm compressor in the air compressor test bed

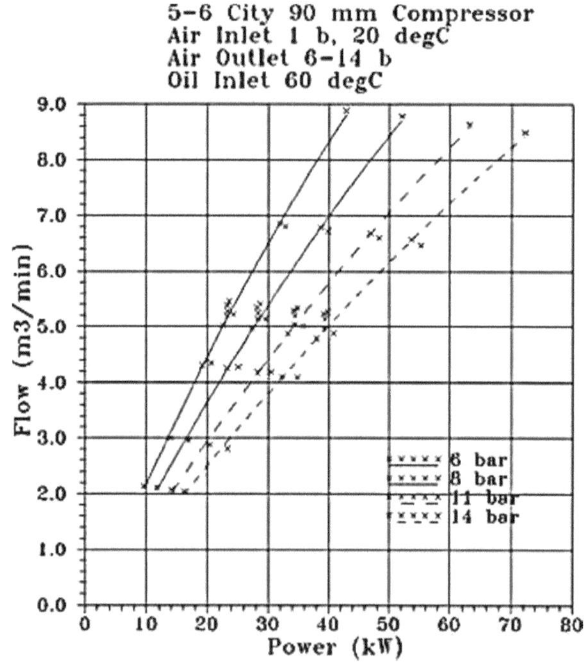

Fig. 4.9. Comparison of the estimated and measured flow as a function of compressor power

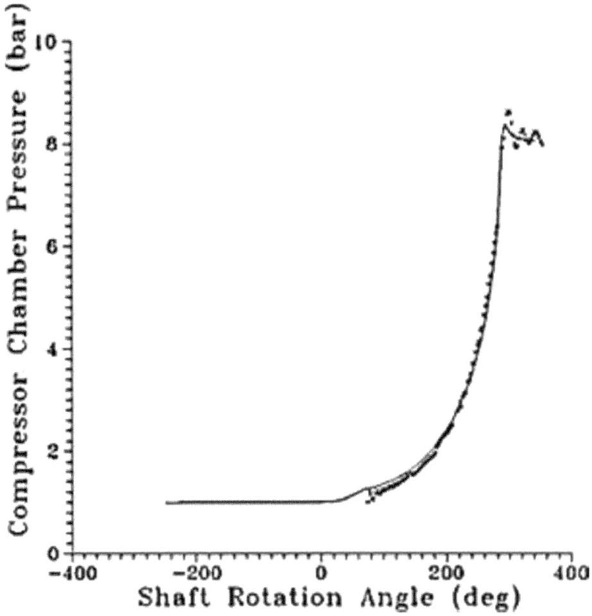

Fig. 4.10. Comparison of the estimated and measured pressure in the compression chamber

located in the bores of the compressor casing, which were logged by means of an oscilloscope.

These and other measurements of temperature and pressure, taken at the compressor inlet and discharge were recorded in a computerised data logger to be processed and displayed in real time. A sample screen record of this is given in Fig. 4.7.

Test results, corresponding to the design point showed a measured air flow rate of 5.51 m^3/min, compared to a calculated value of 5.56 m^3/min, while the measured power input was 32 kW, compared to a calculated value of 31.7 kW. Other results are shown graphically in Fig. 4.9, while a comparison between the measured and estimated pressure change within the working chamber is shown in Fig. 4.10.

5

Examples of Modern Screw Compressor Designs

The study described in Sect. 4.2 shows some of the advantages which can be obtained by the adoption of computer aided design techniques, developed from fundamental equations of heat and fluid flow and the application of the mathematical theory of gearing to rotor design. Publications showing awareness of the advantages of such an approach are exemplified by Fleming and Tang, 1994, 1998, Stosic and Hanjalic, 1997, Xing et al., 2000, and Kovacevic et al., 2004.

More recently, the authors have been and are continuing to make efforts to integrate the various independent considerations required to carry out a comprehensive computer aided design in an interactive manner. This also makes provision for the inclusion of additional analytical packages for more advanced analyses, as they are developed. The aim is to develop a design methodology to proceed from specification of only a limited number of input parameters to working drawings, in the minimum number of iterative steps. Central to this is a design interface for screw compressors which links the various segments and cross references any changes made in one set of routines with the others. It thus serves as an envelope, which interfaces with the following subroutine packages:

A suite of computer programs for screw compressor optimal rotor profiling and thermodynamics carries out the rotor profiling, and thermodynamic and fluid flow analysis, based on one dimensional considerations using a multi variable minimisation procedure to obtain the best result. Simultaneous variation of up to nine variables for a single stage machine and eighteen variables in a two stage compressor have been successfully managed by this means.

Screw compressor rotor grid generator creates a grid for 3-D analysis of flow in the compressor.

A commercial software package for 3-D analysis of fluid flow and fluid-solid interaction effects is used to solve transport equations of flow and elastoplastic body dynamics.

Rolling element bearing data is used for bearing selection in the design process.

A CAD package for 2-D or 3-D presentation of the derived design details.

Such an approach to design is used by only a very limited number of companies and then only in part. The following sections give some details of a selection of applications designed completely, or in part, by the authors, which have benefited from the use of these software routines. All of those described, have been built as prototypes or production machines. However, as already indicated, where applicable, the use of such software has always been accompanied by careful attention to detail design of the bearings, porting and all items which may affect the ultimate performance of a twin screw machine, whether acting as a compressor or an expander. In addition, some examples of machines are given in which a single rotor pair has been used to perform two functions simultaneously. In all cases the designs were based on rack generated "N" profile rotors.

5.1 Design of an Oil-Free Screw Compressor Based on 3-5 "N" Rotors

This section briefly describes two oil-free compressors designed for dry air delivery. A 3/5 rotor configuration was chosen, as shown in Fig. 5.1. This arrangement gives a larger cross section area with stronger gate lobes than any other known screw compressor rotor. Also, by using the timing gears to transfer the torque, the main rotor speed was increased by 1.67:1. Advantage was taken of this high step-up ratio to enable a single stage gearbox to be used for connecting the compressor to the drive shaft, whereas for this application, a two-stage gearbox is normally required.

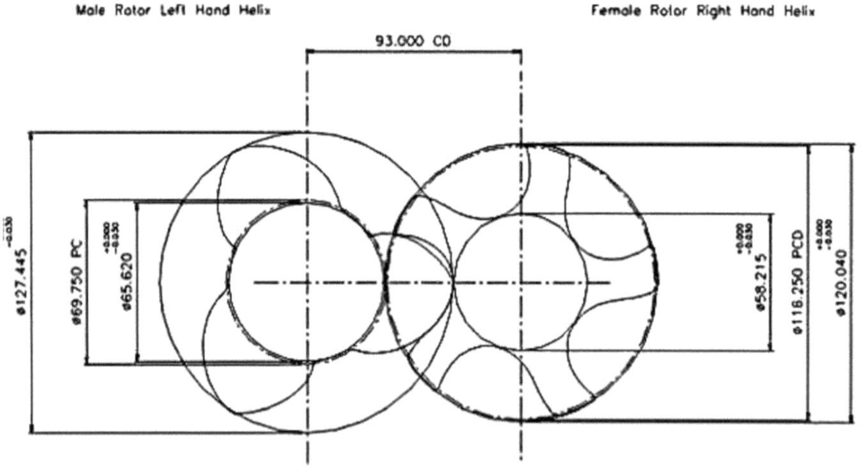

Fig. 5.1. Drawing of 3/5 "N" rotors used in XK12 and XK18 compressors

Fig. 5.2. Photograph of 3/5 "N" Rotors used in XK12 and XK18 compressors

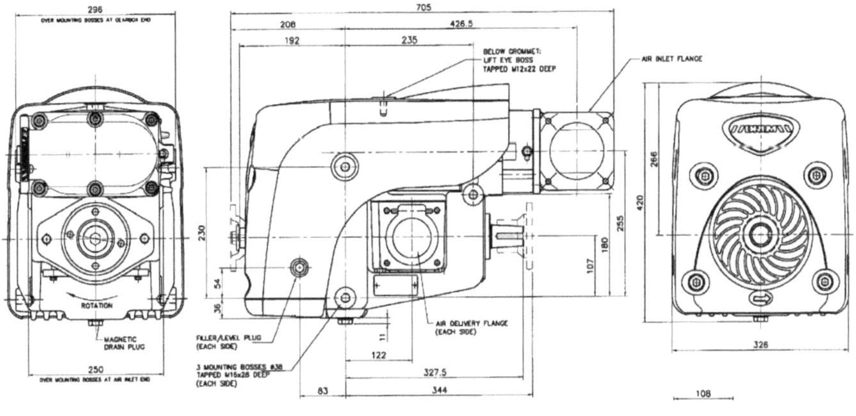

Fig. 5.3. Outline dimensions of the XK12 compressor

Two sizes of compressors were designed to cover a delivery range of 300 to 1100 m^3/hr. Compressor XK18 was designed for a nominal 1000 m^3/hr and XK12 for 500 m^3/hr.

The design was carried out jointly with industry so that the rotors and porting were designed by the authors. A photograph of the rotors is given in Fig. 5.2. The remainder of the mechanical design was done by the manufacturers. An outline drawing of the XK12 is shown in Fig. 5.3 and a photograph of the XK18 in Fig. 5.4.

Fig. 5.4. Photograph of the XK18 compressor

Both compressors have a gearbox step-up ratio of 4.2:1. This gives an overall step-up ratio of 7:1 when combined with the synchronising gears. The drive shaft speed range is 1000–1800 rpm. Thus the main rotor speed is in the range of 7000–12600 rpm. The gear train and discharge bearings are oil lubricated, while all other bearings are grease lubricated.

XK12 and XK18 overall dimensions and weight are 617 × 300 × 390 mm, 113 kg and 700 × 300 × 390 mm, 127 kg respectively. A through shaft gearbox arrangement allows a clockwise or anticlockwise input drive rotation with mounting points on either side of the machine to provide further flexibility. The inlet manifold designs available can accommodate application access to the axial inlet port from either left, right, vertical or the axial direction. The compressor discharge exists from either side of the compressor.

Finally, in Fig. 5.5 the performance of these compressors at discharge pressure of 3 bars is compared with the reference compressor R2, which is the D-9000 made by the same manufacturer, R1, which is C-80 of GHH based on SRM "A" profile rotors. The latter, despite its age outperformed other reference compressors, such as compressors R3, which is the Typhoon made by Mouvex, based on modern screw compressor profiles and R4, which is the GHH CS1000. This is also based on the SRM "A" profile.

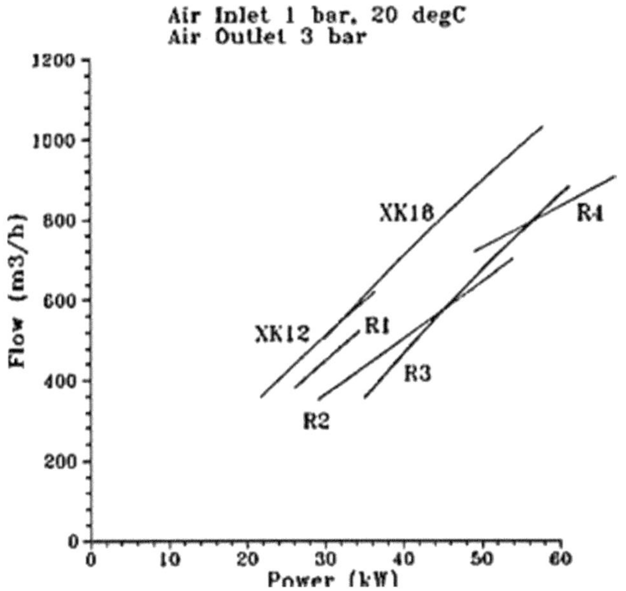

Fig. 5.5. Flow as a function of power for the XK12 and XK18 compared with reference compressors

As can be seen, the flow of the XK12 and XK18 compressors, which is actually greater than the predicted value, is at least 10% higher for the same compressor power than any of its competitors.

5.2 The Design of Family of Oil-Flooded Screw Compressors Based on 4-5 "N" Rotors

The majority of screw compressors are still manufactured with 4 lobes in the main rotor and 6 lobes in the gate rotor with both rotors of the same outer diameter. This configuration is a compromise which has favourable features for both, dry and oil-flooded compressor application and is used for air and refrigeration or process gas compressors. However, other configurations, like 5/6 and 5/7 and recently 4/5 and 3/5 are becoming increasingly popular. Five lobes in the main rotor are suitable for higher compressor pressure ratios, especially if combined with larger helix angles. The 4/5 arrangment has emerged as the best combination for oil-flooded applications of moderate pressure ratios and its configuration permits the smallest overall rotor dimension than that for any other reasonable combination. Also, one less lobe in the gate rotor compared with the 4/6 combination can improve the efficiency of rotor manufacture.

On this basis, a family of oil flooded air compressors was designed by the authors in conjunction with industry based on the use of rack generated "N" profiles in a 4/5 configuration. The range of delivered flows required was from 0.6 to 60 m^3/min, at delivery pressures of 5–13 bars gauge. Using the same L/D ratio of 1.55:1 throughout, this was achieved using only 5 rotor sizes, namely; 73, 102, 159, 225 and 284 mm main rotor diameter. An end view of the 102 mm rotor pair is shown in Fig. 5.6 and a photograph of a finished pair is given in Fig. 5.7.

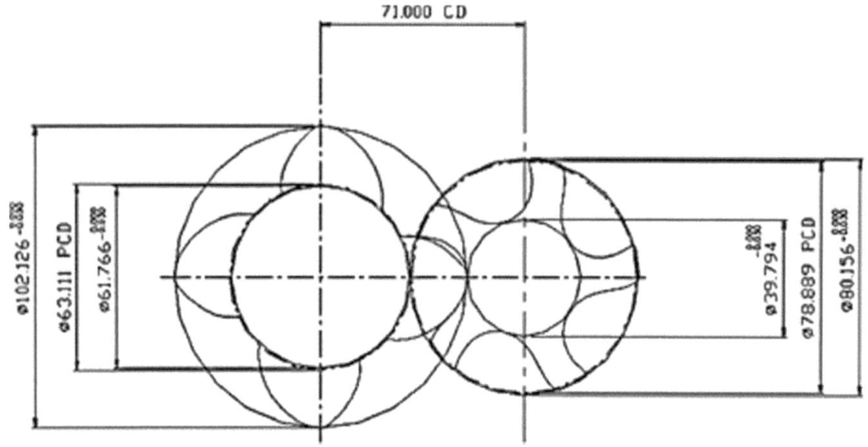

Fig. 5.6. 4/5 "N" rotors for the family of screw air compressors

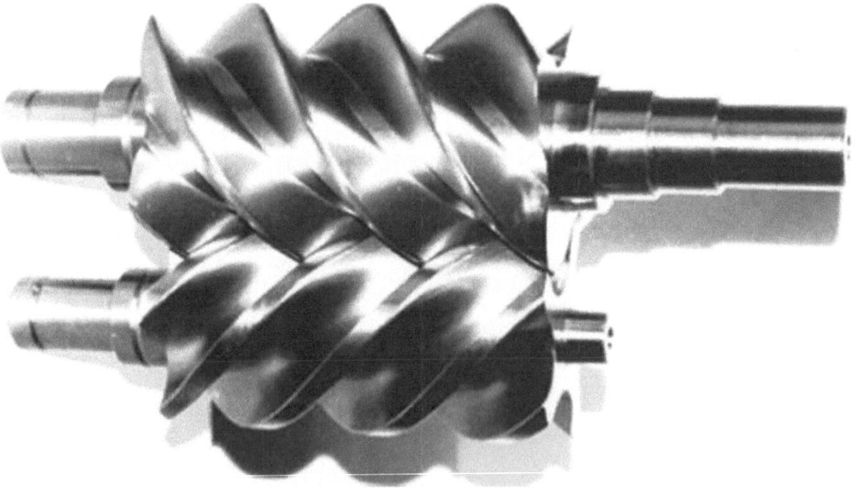

Fig. 5.7. Photograph of 4-5-102 mm "N" rotors

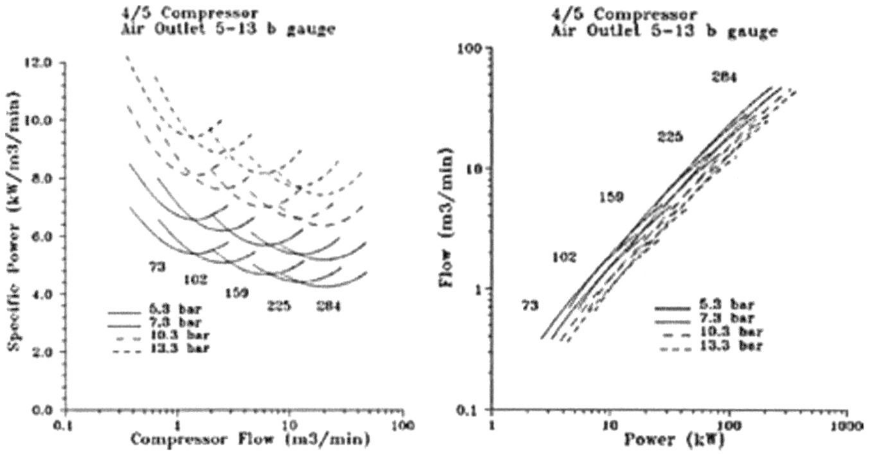

Fig. 5.8. Predicted Performance of the Elgi family of oil-flooded air compressors

The main performance parameters calculated by the simulation program, on which the designs are based, are presented in logarithmic coordinates in Fig. 5.8.

The mechanical design of the compressor family was performed interactively and involved close collaboration between the authors and the manufacturer's own engineers. Thus, the first two compressors were designed by the authors, the third and fourth by the manufacturer's own designers under direct supervision of the authors, while the fifth compressor was designed entirely by the manufacturer's engineers. All activities related to tool and rotor manufacture and complete machine production and assembly were performed at the manufacturer's premises in India. A cross sectional view and photographs of the completed 73 mm compressor are given in Fig. 5.9.

The prototype 73 and 102 mm compressors were tested by the authors and a view of the 73 mm machine, mounted on the test rig, is shown in Fig. 5.10, while test results for this machine are given in Fig. 5.11.

As may be seen from Fig. 5.11 the correlation between the estimated and measured performance is good. These results, together with those obtained from other machines in the family, were compared with performance details of equivalent compressors, as published in the brochures of a number of well known manufacturers. It was concluded that not only were the efficiencies of the authors compressors higher than the majority of these but the "N" profile family of compressors was all also substantially smaller and lighter.

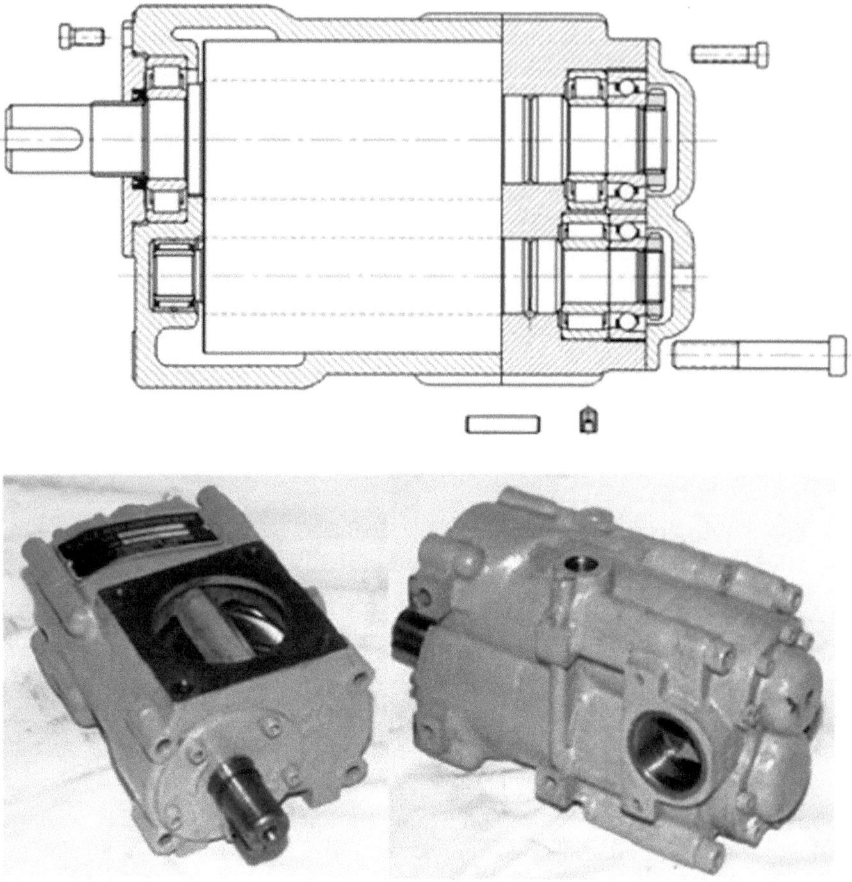

Fig. 5.9. Smallest compressor of the family, 4/5-73 mm

5.3 Design of Replacement Rotors for Oil-Flooded Compressors

The market for screw compressors is highly competitive, especially in compressed air and refrigeration systems, and new designs are continually being introduced which are more efficient and cost effective than their predecessors. However, because of the high cost of development of new machines, manufacturers seek to maintain their existing designs for as long as possible. Closer study of many of the older designs has shown that in the majority of cases, all that is required to bring them up to date is to change the rotor profile to one of more up to date type.

An opportunity arose to try this on a family of highly efficient oil-flooded air compressors. This comprised four machines with rotor diameters of 102,

Fig. 5.10. Compressor test layout

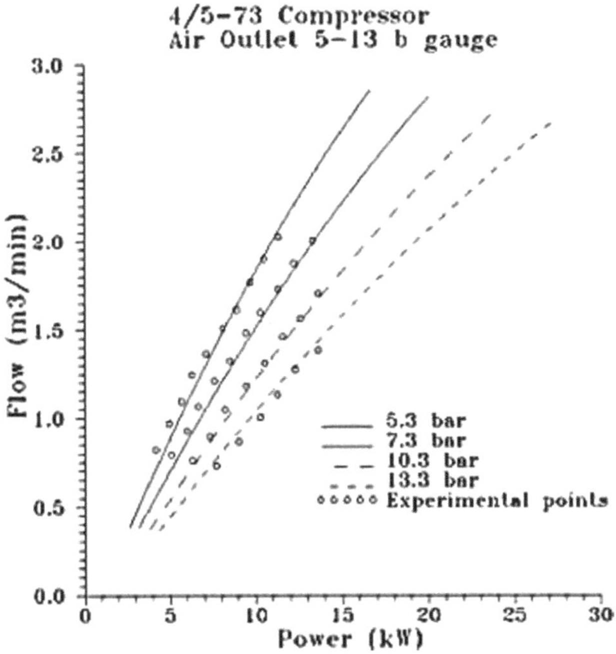

Fig. 5.11. Comparison of the estimated and measured compressor data

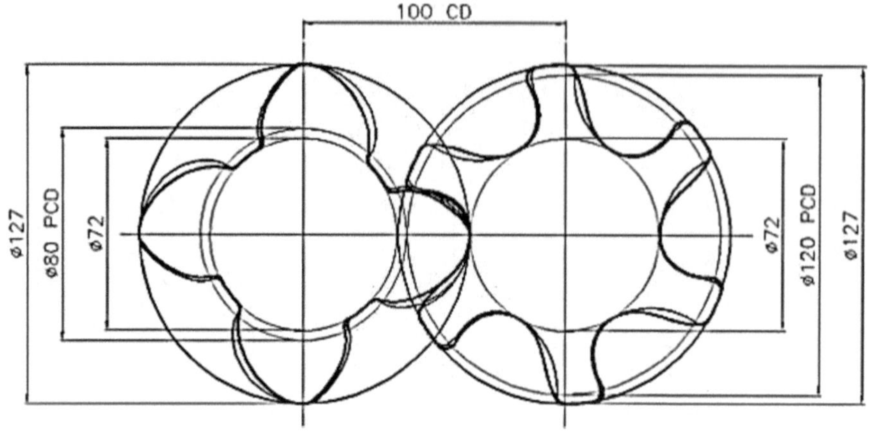

Fig. 5.12. "N" Rotor Replacement (*bold*) for Asymmetric Rotors (*light*)

Fig. 5.13. Comparison of asymmetric rotors (*left*) and "N" rotor replacement (*right*)

127, 163 and 204 mm all fitted with standard asymmetric rotors with a 4/6 configuration.

The design software package, already described, was used to design a replacement set of rotors based on the rack generated "N" profile. The same 4/6 configuration was adopted and all the definitive dimensions of the original rotors, such as centre distance, outer and root diameters and rotor length were retained. Due to the favourable features of the "N" profile, it was possible to design and manufacture the replacement rotors with smaller clearances than those of the original at no extra cost and thereby effect further performance improvements. However, to make a fair comparison, the clearances adopted were identical to those of the original rotors.

Figure 5.12 is a drawing of the new and old rotors of the 127 mm compressor, while a photograph of them, manufactured is given in Fig. 5.13. The new rotors had a 5.4% greater displacement and a 4.5% longer sealing line than the originals. Better torque distribution on the new rotors reduced the contact

Fig. 5.14. Compressor with replacement rotors installed

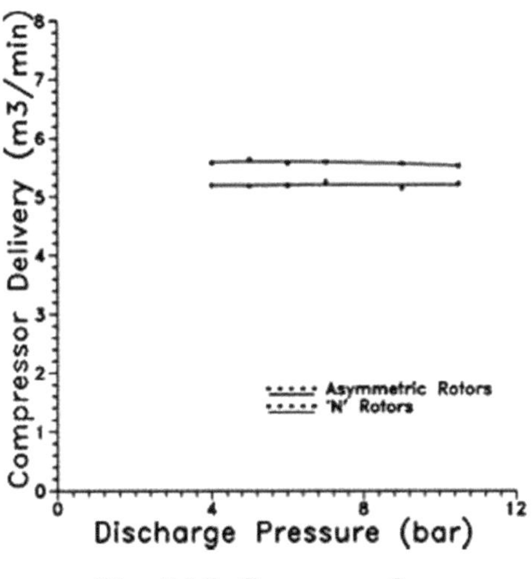

Fig. 5.15. Compressor flow

stress between the main and gate rotors and minimised the chance of their rattling. An overall view of the assembled compressor is shown in Fig. 5.14.

Comparative test results of the compressor with the asymmetric and "N" profile rotors are given in Figs. 5.15 to 5.17. As can be seen, the new rotors increased the compressor delivery rate by 6.5% while decreasing the minimum specific power by at least 2.5%.

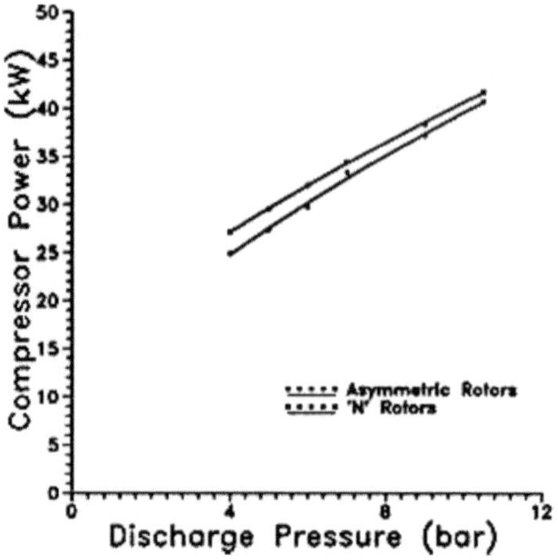

Fig. 5.16. Compressor power

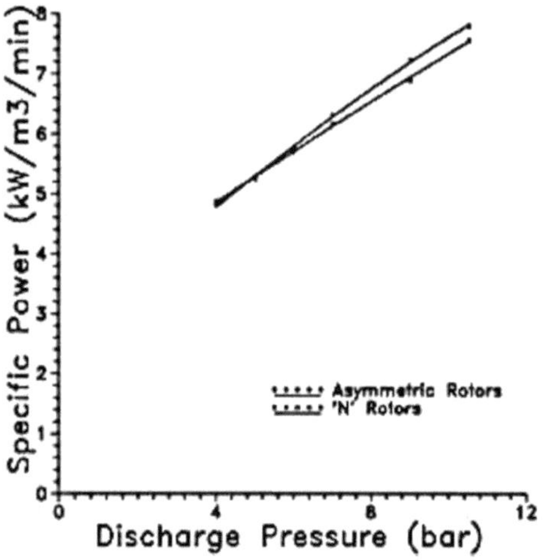

Fig. 5.17. Specific power

5.4 Design of Refrigeration Compressors

Screw compressors used for refrigeration systems are usually of the semi hermetic type, as shown in Fig. 5.18. Due to increasingly stringent environmental protection protocols and the development of new refrigerants to meet them, refrigeration compressor designers continually have to redesign these machines to allow for new sizes, built in volume ratios and pressure differences, as well as problems associated with fluid viscosity and fluid-oil miscibility, all of which arise from the use of these fluids. Moreover, there is a continuing need to make refrigeration compressors to operate more quietly, at higher efficiencies and with a longer service life.

Fig. 5.18. Typical semihermetic refrigeration screw compressor nested within oil separator with a slide valve capacity control

These factors all lead to the need to optimise compressor designs as comprehensively as possible, before construction and testing. A further trend has been to seek out innovative ideas to bring basic improvements to the entire design concept. Some of these are not necessarily new but have become worthwhile or feasible in recent years as a result of the use of the new fluids, improved manufacturing ability and condenser and evaporator designs having reached a level where their further improvement becomes too expensive. One approach, which has been considered in detail by the authors is to use a single pair of rotors for more than one function such as two stage compression or combining compression with expansion. These considerations are reviewed in more detail in later sections.

5.4.1 Optimisation of Screw Compressors for Refrigeration

As shown in earlier sections, even a simple analysis of rotor behaviour in refrigeration compressors shows that a number of desirable rotor characteristics lead to conflicting design requirements. This implies that, given the compressor duty, simultaneous optimisation of all the variables involved in the design process must be performed to obtain the best possible performance.

Minimization of the output from the process equations leads to the optimum screw compressor geometry and operating conditions. These can be defined as either the highest flow and compressor volumetric and adiabatic efficiencies, or the lowest compressor specific power. More information on screw compressor optimisation is given by Stosic et al., 2003b, where an example is given involving nine variables. These include the rotor radii, defined by four rotor profile parameters, the built–in volume ratio, the compressor speed and the oil flow, temperature and injection position.

A variety of similarly optimized rotors and compressors, all of which are used by the compressor industry, are presented throughout this and the next section. Figure 5.19 shows one of them, when rotors with unequal diameters are used to minimize the blow-hole area to give increased displacement and better efficiency when compared to a design with equal diameter rotors.

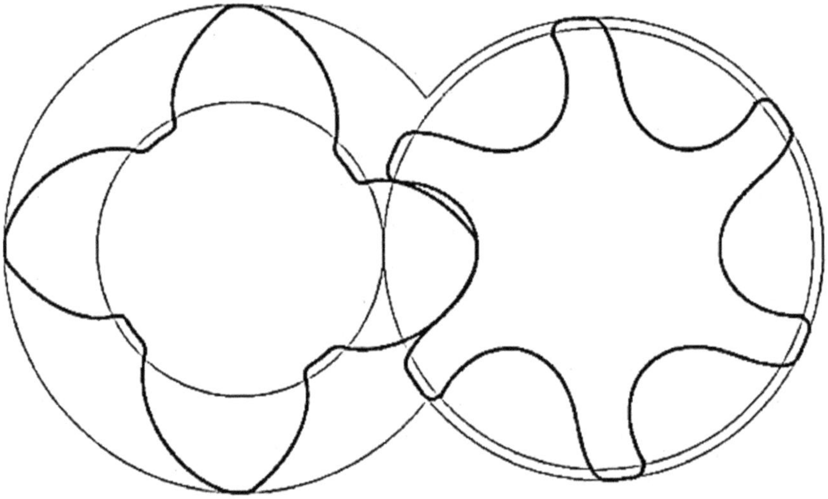

Fig. 5.19. Rotors optimized for best efficiency, unequal rotor diameter

5.4.2 Use of New Rotor Profiles

The need to improve compressor capacity and performance has been the incentive for many screw refrigeration compressor manufacturers to adopt more modern profiles, especially since many of the earlier designs were developed for air compressors. Optimisation was then employed to obtain the best delivery and highest efficiency for the same rotor tip speed. Rigorously applied, this leads to the need for a different rotor design for each application.

Two different rotor pairs are shown in Fig. 5.20. One of these is for light refrigeration duty and air conditioning, where the pressure ratios and pressure differences are low to moderate. The other is for heavy refrigeration duty where both pressure ratios and differences are relatively high.

A somewhat more general design, which will perform reasonably well at both high and low pressure ratio and differences, is shown in Fig. 2.21.

5.4.3 Rotor Retrofits

When tooling costs are large and users are generally satisfied with their product, compressor manufacturers are reluctant to modify their compressor housings and systems. However, if the rotor configuration, centre distance and outer rotor diameters are kept the same, all other profile parameters can be modified to achieve both better capacity and efficiency.

An example of this is a 4/6-204 mm rotor retrofit in an ammonia compressor, shown in Fig. 5.22. The retrofit rotors had about 5% larger displacement and their blow-hole area was about 20% less. Due to low torque on the gate rotor, the rotor contact force was small, which resulted in a lower mechanical loss due to friction between the rotors. As a result a better compressor performance was achieved for the retrofit rotors in comparison with the previous ones. The experimental results, obtained at the manufacturer's site, are shown in the Table 5.1.

5.4.4 Motor Cooling Through the Superfeed Port in Semihermetic Compressors

It is well known that motor cooling by the compressor suction gas introduces capacity and efficiency loss penalties. Both of these effects are caused by the temperature rise of the cooling gas resulting from cooling the motor. To overcome that problem, superfeed vapour can be used instead and injected into the motor housing. That keeps the compressor and plant capacity and the compressor efficiency virtually the same as for an open compressor. The same compressor may be used even if the superfeed is not present in the plant, when a portion of high pressure liquid can be injected for the motor cooling instead of the superfeed vapour.

If the plant has no superfeed, a somewhat similar effect can be achieved if high pressure liquid is used for that purpose and the vapour resulting from

Fig. 5.20. Screw compressor rotors optimized for air conditioning and light refrigeration duty, *left* and rotors designed for heavy refrigeration duty, *right*

the motor cooling process is evacuated through the superfeed port. Apart from the inevitable resulting decrease in the plant capacity, the compressor efficiency will be unchanged. A compressor with such a cooling concept is shown in Fig. 5.23.

5.4.5 Multirotor Screw Compressors

The use of multiple main or gate rotors in one screw compressor to increase capacity was proposed almost at the introduction of these machines. In Fig. 5.24

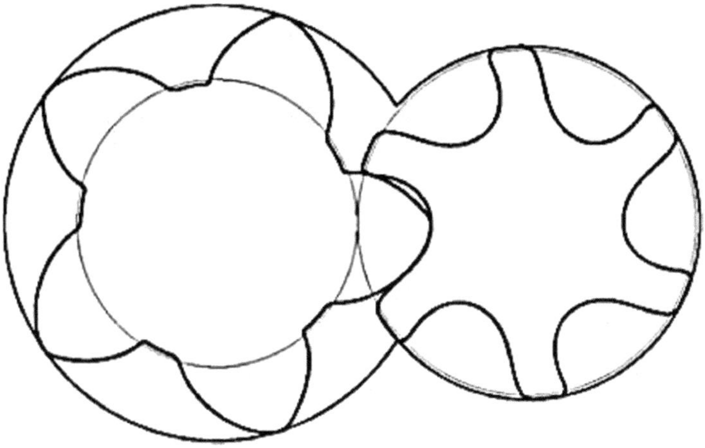

Fig. 5.21. Original rotors and compressor optimized for general refrigeration and air conditioning duty

Table 5.1. Experimental Comparison of Compressor Performance with Retrofit and Standard Rotors

Standard Rotors			
Evaporation/Condensation Temp	−15/30°C	−35/35°C	
Shaft Speed [rpm]	2920	2920	
Refrig Capacity [kW]	626	216	
Motor Power [kW]	178	156	
COP	3.523	1.383	
Optimized Rotors			
Evaporation/Condensation Temp	−15/30°C	−35/35°C	0/35°C
Shaft Speed [rpm]	2920	2920	2920
Refrig Capacity [kW]	669	243	1187
Motor Power [kW]	182	168	245
COP	3.671	1.486	4.98
COP Improvement New/Old	104.2%	107.5%	-

a multirotor screw compressor with two gate rotors is shown, as given in Sakun, 1960. The idea has not yet been fully commercialised. However, several patents, such as those of Shaw, 1999 and Zhong, 2002 have been recently published in that area. It is obvious that the capacity of a multirotor compressor will be a multiple the capacity of the corresponding ordinary screw compressor. Nonetheless, although it is fairly self evident, it is not yet fully appreciated that the efficiency of a multirotor machine will be no better than that of a number of single rotor pair compressors.

Another feature of the multirotor arrangement is the balancing of radial forces on the main rotor. Unfortunately the axial forces on the main rotor

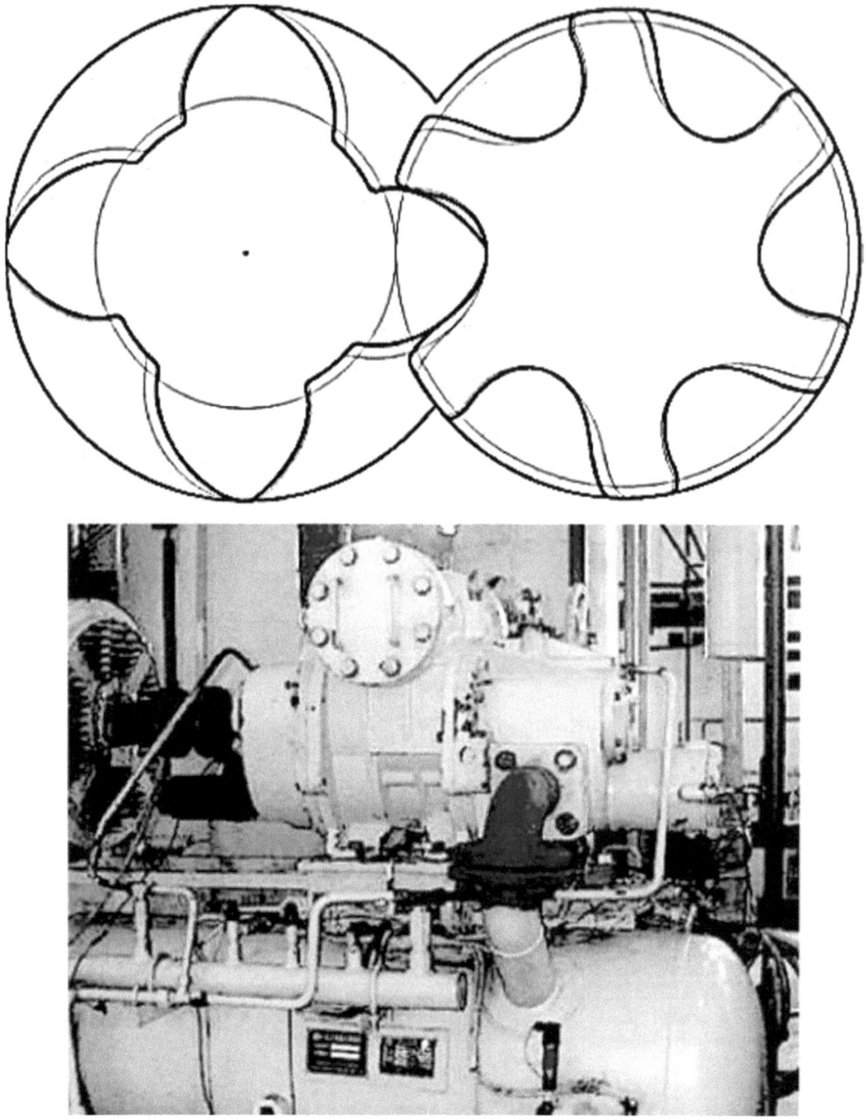

Fig. 5.22. Compressor and retrofit rotors optimized for general refrigeration duty compared with the original rotors (*light line*)

Fig. 5.23. Semihermetic compressor with motor cooling through superfeed port at the motor housing, far right

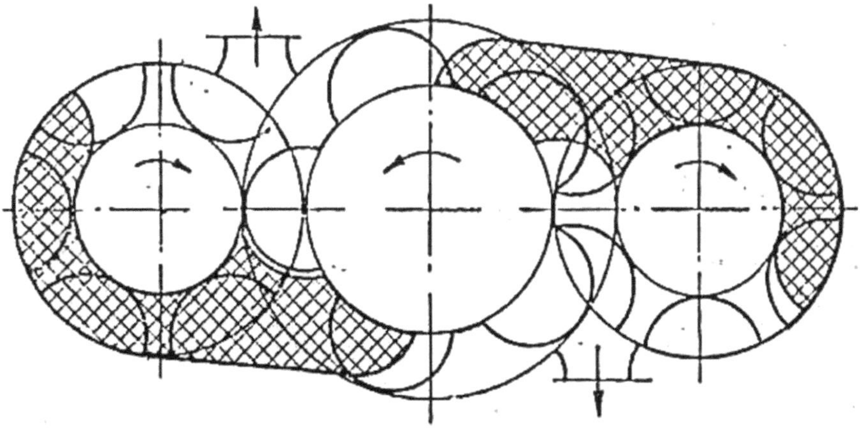

Fig. 5.24. Layout of the multirotor screw compressor

are simultaneously multiplied. Generally, it is easier to cope with axial than with radial rotor forces by using, for example balancing pistons. Hence this feature can be regarded as an advantage. The gate rotor forces are virtually unaffected by this arrangement.

5.5 Multifunctional Screw Machines

One of the potential advantages of screw machines over other types of positive displacement machine is their ability to perform both the compression and expansion functions simultaneously, using only one pair of rotors. A further feature of this is the use of the rotors which seal on both contacting surfaces so that the same profile may be used both for the expander and the compressor sections. This means the compressor and expander rotors can be machined or ground in a single cutting operation and then separated by machining a parting slot in them on completion of the lobe formation. Moreover, by location of the machine ports, as shown, in Fig. 5.25, the pressure loads can be partially balanced and thereby, mechanical friction losses will be less than if the two functions are performed in separate machines.

5.5.1 Simultaneous Compression and Expansion on One Pair of Rotors

Fields of application of such machines are replacement of the throttle valve in refrigeration and air conditioning plants, high pressure applications, fuel cells, multistage compression or expansion and, really, any other application where simultaneous compression and expansion are required. One example of such an unusual, but convenient application is compressor capacity control by partial expansion of the compressed gas.

As is shown in Fig. 5.25 high pressure gas enters the expander port at the top of the casing, near the centre, and is expelled from the low pressure port at the bottom of the casing at one end. The expansion process causes the temperature to drop. However, here the fall in pressure is used to recover power and causes the rotors to turn. Gas enters the low pressure compressor port, at the top of the opposite end of the casing, is compressed within it and expelled from the high pressure discharge port at the bottom of the casing, near the centre. Ideally, there is no internal transfer of fluid within the machine between the expansion and compression sections which each take place in separate chambers.

If the same machine presented in Fig. 5.25 is used as a two stage compressor, only the ports of the second stage will exchange their places. The low pressure port of the second stage will be located on the top of the machine and the high pressure discharge will be at the machine bottom. This offers a compact two stage machine which may be used either in the oil flooded or dry operation mode. A similar arrangement is valid for a two stage expander.

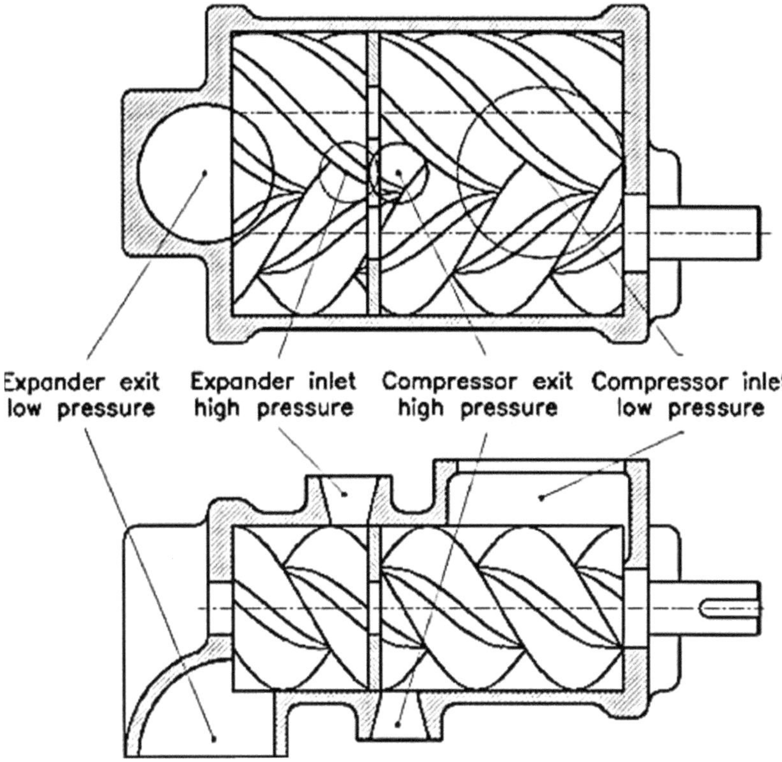

Fig. 5.25. View of the multifunctional rotors acting simultaneously as compressor and expander

5.5.2 Design Characteristics of Multifunctional Screw Rotors

Since compression and expansion are carried out separately in multifunctional rotors the compressor and expander profiles could be different. However, this would make manufacture extremely difficult, due to the small clearance space between the two rotor functions. Hence to make it possible for the proposed multifunctional rotors to be utilized, the rotors must form a full sealing line on both contacting surfaces so that the same profile may be used both for the expander and the compressor sections.

An example of how the rotor profile will then appear is given in Fig. 5.26.

Additionally the expansion section can contain a capacity control such as a slide or lifting valve at suction to alter the volume passing through it at part load, in a manner identical to capacity controls normally used in screw compressors.

If the rotors are used for multistage compression, they can retain their profile shape common for screw compressors with a small blow hole on one side and a relatively large one on the opposite side.

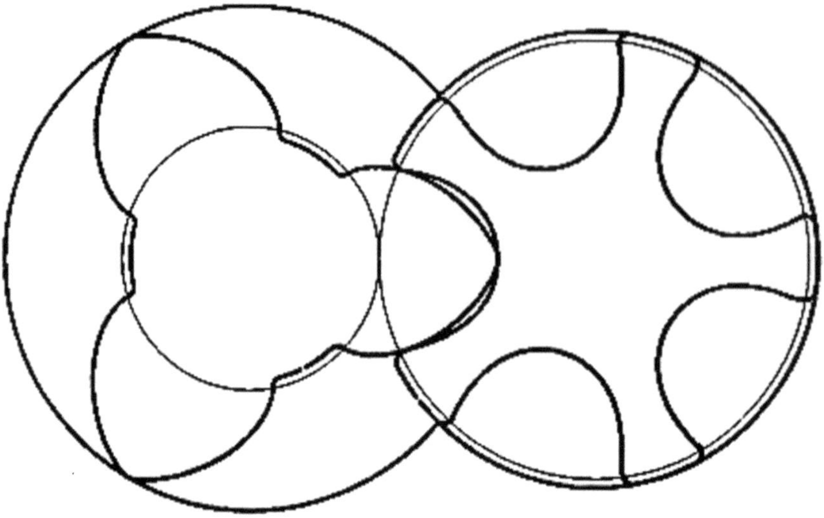

Fig. 5.26. Compressor-Expander Rotors

5.5.3 Balancing Forces on Compressor-Expander Rotors

An important novelty of the compressor expander arrangement on one pair of rotors is in the positioning of the ports. Because the high pressure ports of such machine are in the centre of the unit and arranged so that they are on opposite sides of the casing, the high pressure forces due to compression and expansion are opposed to each other and, more significantly, only displaced axially from each other by a relatively short distance. The radial forces on the bearings are thereby significantly reduced. In addition, since both ends of the rotors are at more or less equal pressure, the axial forces virtually balance out The following example of a combined compressor and expander in the high pressure application indicates the extent of the advantages, which are possible from this arrangement.

A refrigerator uses $2.75\,\mathrm{m}^3/\mathrm{min}$ CO_2 as a working fluid which leaves the evaporator and enters the compressor as dry saturated vapour at a suction pressure of 35 bar to leave the compressor and enter the condenser at a discharge pressure of 100 bar. The compressor rotor required would be 102 mm in diameter with a length/diameter ratio of 1.5. The expander required to replace a throttle valve in this system would have a main rotor of the same diameter but with a length/diameter ratio of only 1.1. Force calculations showed what bearing loads must be resisted if the refrigeration system is designed with a conventional screw compressor drive. On the main rotor alone, there is an axial force of 92 kN and radial bearing forces of 132.9 kN at the high pressure end and 45,5 kN at the suction end. A similar calculation was performed for the expander rotors and their corresponding bearing forces. Here, the axial bearing load on the main rotor is 91,9 kN while the corresponding radial loads are 85.9 kN at the high pressure end and 34,1 kN at the low pressure end. The bearing forces, which would result, if the compressor and expander rotors were machined on the same shafts with the high pressure ports in the middle and the low pressure ports at each end were as following. The main rotor axial load has been reduced to 0.12 kN, which is negligible. The radial bearing loads are now 101 kN at the compressor end and 117 kN at the expander end. More significantly, for the gate rotor, which is weaker, the maximum bearing load has been reduced from 146 kN to 119 kN, which is 19% less. Thus the total bearing load on the main rotor alone has been reduced from 270.4 kN for the compressor to 218 kN for the combined compressor-expander. If both main and gate rotors are included, then the total bearing load is reduced from 556 kN for the compressor alone to only 448 kN for the combined balanced rotors. This amounts to a total decrease in bearing load of nearly 20%. Design problems associated with high bearing loads in screw compressors for CO_2 systems are thereby reduced.

5.5.4 Examples of Multifunctional Screw Machines

Several examples of application of multifunctional rotors are presented here. The expressor for simultaneous expansion of refrigerant liquid and compression of its vapour, a dry fuel cell compressor expander for simultaneous compression of air and expansion of fuel cell reaction products, a high pressure oil flooded compressor for CO_2 and a two stage oil flooded air compressor.

The Expressor

An introductory report on a means of replacing the throttle valve in vapour compression systems was published by Brasz et al., 2000. Power is recovered from the two-phase expansion process and used directly to recompress a portion of the vapour formed during the expansion. Both the expansion and recompression processes are carried out in a self-driven machine with only one pair of rotors and no external drive shaft. The principle of simultaneous expansion and compression on the same pair of screw rotors is illustrated in Fig. 5.27. Unlike rotors that perform compression or expansion only, these have to seal on both sides, as shown in Fig. 5.28. The authors have called such a device an "expressor" and, as built and tested, it operated as a process lubricated totally sealed unit without the need for lubricating oil, internal seals or timing gears. A complete unit is shown in Fig. 5.29.

The test results indicate that the overall expansion-compression efficiency of the expressor is of the order of 55%. This corresponds roughly to 70% expansion efficiency and 80% compression efficiency. The simplicity of the expressor design, together with its promising performance, indicate that it should be a highly cost effective component in large commercial chiller systems.

Further studies are being performed to determine the best built in volume ratios for the expansion and compression processes and to improve liquid-vapour separation during low pressure discharge.

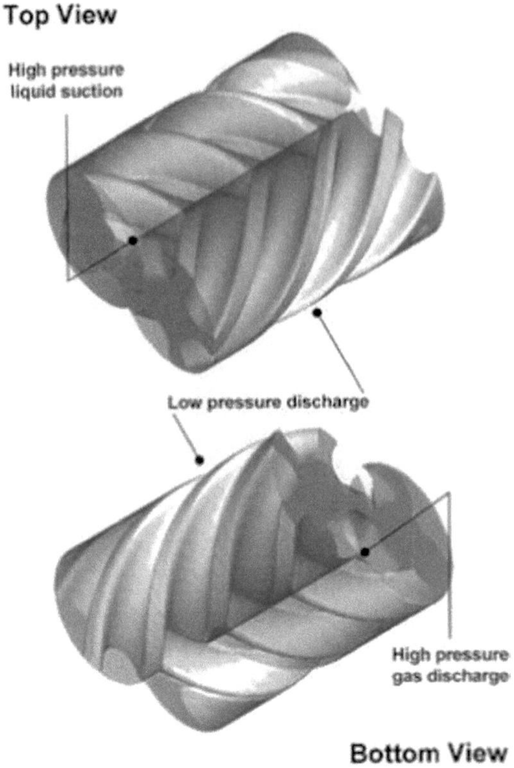

Fig. 5.27. Expressor Principle

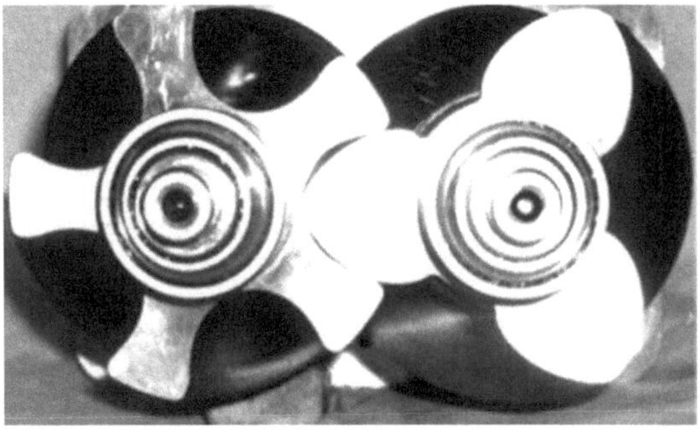

Fig. 5.28. Expressor Rotors Sealed on Both Sides

Fuel Cell Compressor-Expander

An expanded view of the compressor expander for fuel cell application is given in Fig. 5.30. Both, the expansion and compression sections are clearly visible. These are separated by the central plate which contains the high pressure ports.

Fig. 5.29. The Expressor Prototype

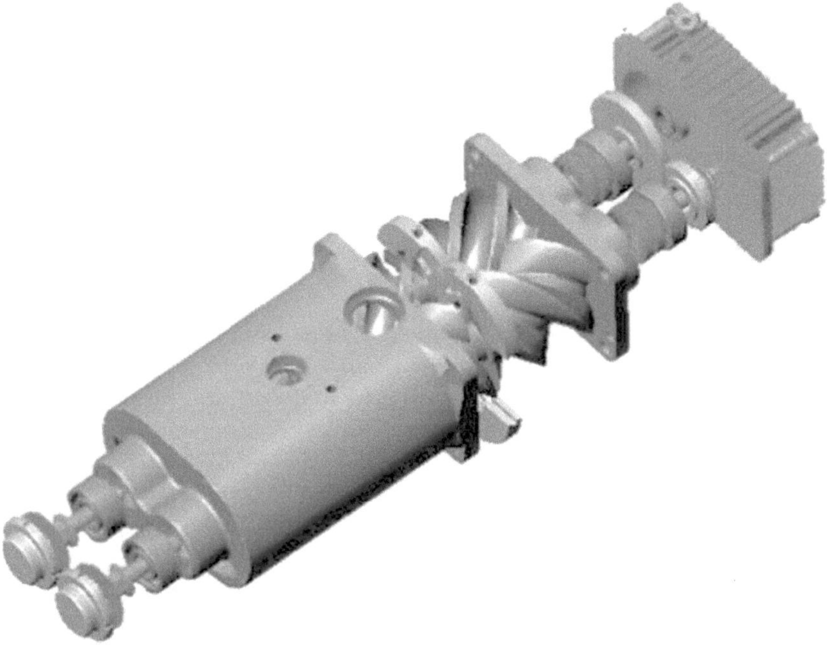

Fig. 5.30. Expanded View of the Compressor-Expander for Fuel Cell Application

Fig. 5.31. Expanded View of the Compressor-Expander for Fuel Cell Application

A prototype of a similar machine was manufactured and experimentally investigated as shown together with its drive motor in Fig. 5.31.

High Pressure Screw Compressor

Recent interest in natural refrigerants, has resulted in more intensive studies of CO_2 as a working fluid in vapour compression systems for refrigeration and air conditioning. Two major drawbacks to its use are the very high pressure differences of up to 50 bars required across the compressor and the large efficiency losses associated with the throttling process. To overcome the throttle losses, combined compression with recovery of work from the expansion process is proposed and, as described in Sect. 5.5.3, the bearing loads are thereby reduced.

An analysis presented by Stosic, 2002 shows that the coefficient of performance will be improved by both recovery of the throttle loss and reduction of the mechanical losses because of lower rotor loads and thus be increased by 72% from 2.79 to a more acceptable 4.8. However, these figures are based on idealised work input and output. In a practical system, allowance must be made for the compression and expansion efficiencies, which would reduce the expansion work and increase the compression work. Nonetheless, an overall gain in coefficient of performance over the ideal cycle with a throttle valve

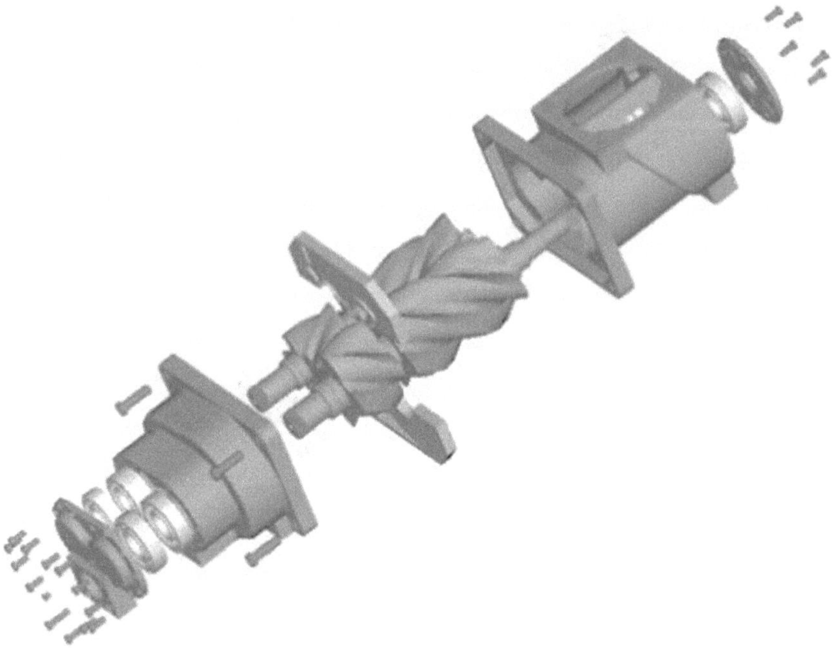

Fig. 5.32. Expanded View of the two-stage compressor on one pair of rotors

should still be achievable by this means. More information on this application can be found in Stosic et al., 2002.

Two-Stage Compressor

Another prospective application of multifunctional rotors is for two stage compression or expansion. Namely, one pair of screw rotors can be used for simultaneous compression in both the first and second compressor stages. Such a compressor is presented in Fig. 5.32. As can be seen in the figure the rotor diameter, as well as the rotor profiles are the same. The first stage is on the right side and its rotors are longer compared with the second stage, which is presented on the left. The principle has been known for some time, but have been no successful applications reported. A revival of the concept was initiated by Varadaraj, 2002. A two stage compressor on one pair of rotors has been designed and compressor prototypes are being manufactured.

Capacity Control by Expansion of Surplus Compressed Gas

An interesting application of multifunctional rotors is proposed here. This is a means of compressor capacity control by expanding the flow surplus in the expander part of the combined compressor expander. It is known that screw compressors are subjected to various capacity controls, including suction throttling, variable compressor speed and variable compressor suction volume. Throttling is inefficient, while variable speed devices and sliding valves are expensive, therefore the proposed principle, although burdened by the combined efficiencies of both the compressor and expander process, may still produce benefits.

6

Conclusions

Although the screw compressor is now a well developed product, greater involvement of engineering science in the form of computer modelling and mathematical analysis at the design stage, makes further improvements in efficiency and reduction in size and cost possible. Despite the advanced stage of modern rotor profile generation, there is still scope for new methods or procedures to improve them further and to produce stronger but lighter rotors with higher displacement and lower contact stress. Also, advances in bearing technology and lubrication, must continually be included to obtain the best results. The use of such machines for expansion as well as compression, leads to possibilities of combining both processes in the same machine and thereby extending their range of application. The procedures described in this work serve as a basis for the employment of design software used for the development of screw machines. These require only a few input parameters, which specify the geometry and operating conditions of a screw compressor. These assure interaction of the parametric design process between profile generation, preliminary calculations, 1D and 3D performance calculations and numerical analysis. This allows any changes made in any stage of the design process to be accounted for in all other phases either earlier or later. Therefore, control over the design process is parametrically conducted from only one place, and redundant data and modelling procedures are reduced. This in turn saves both computer resources and time.

A

Envelope Method of Gearing

Following Stosic 1998, screw compressor rotors are treated here as helical gears with nonparallel and nonintersecting, or crossed axes as presented at Fig. A.1. X_{01}, y_{01} and x_{02}, y_{02} are the point coordinates at the end rotor section in the coordinate systems fixed to the main and gate rotors, as is presented in Fig. 1.3. Σ is the rotation angle around the X axes. Rotation of the rotor shaft is the natural rotor movement in its bearings. While the main rotor rotates through angle θ, the gate rotor rotates through angle $\tau = r_{1w}/r_{2w}\theta = z_2/z_1\theta$, where r_w and z are the pitch circle radii and number of rotor lobes respectively. In addition we define external and internal rotor radii: $r_{1e} = r_{1w} + r_1$ and $r_{1i} = r_{1w} - r_0$. The distance between the rotor axes is $C = r_{1w} + r_{2w}$. p is the rotor lead given for unit rotor rotation angle. Indices 1 and 2 relate to the main and gate rotor respectively.

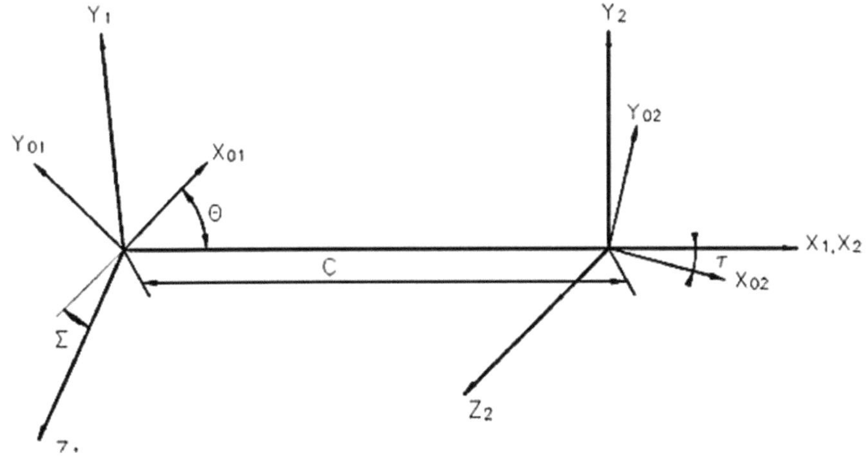

Fig. A.1. Coordinate system of helical gears with nonparallel and nonintersecting axes

The procedure starts with a given, or generating surface $\mathbf{r}_1(t,\theta)$ for which a meshing, or generated surface is to be determined. A family of such generated surfaces is given in parametric form by: $\mathbf{r}_2(t,\theta,\tau)$, where t is a profile parameter while θ and τ are motion parameters.

$$\begin{aligned}\mathbf{r}_1 &= \mathbf{r}_1(t,\theta) = [x_1, y_1, z_1] \\ &= [x_{01}\cos\theta - y_{01}\sin\theta, x_{01}\sin\theta + y_{01}\cos\theta, p_1\theta] \end{aligned} \tag{A.1}$$

$$\begin{aligned}\frac{\partial \mathbf{r}_1}{\partial t} &= \left[\frac{\partial x_1}{\partial t}, \frac{\partial y_1}{\partial t}, 0\right] \\ &= \left[\frac{\partial x_{01}}{\partial t}\cos\theta - \frac{\partial y_{01}}{\partial t}\sin\theta, \frac{\partial x_{01}}{\partial t}\sin\theta + \frac{\partial y_{01}}{\partial t}\cos\theta, 0\right] \end{aligned} \tag{A.2}$$

$$\frac{\partial \mathbf{r}_1}{\partial \theta} = \left[\frac{\partial x_1}{\partial \theta}, \frac{\partial y_1}{\partial \theta}, 0\right] = [-y_{01}, x_{01}, 0] \tag{A.3}$$

$$\begin{aligned}\mathbf{r}_2 &= \mathbf{r}_2(t,\theta,\tau) = [x_2, y_2, z_2] = [x_1 - C, y_1\cos\Sigma - z_1\sin\Sigma, y_1\sin\Sigma \\ &\quad + z_1\cos\Sigma] = [x_{02}\cos\tau - y_{02}\sin\tau, x_{02}\sin\tau + y_{02}\cos\tau, p_2\tau] \end{aligned} \tag{A.4}$$

$$\begin{aligned}\frac{\partial \mathbf{r}_2}{\partial \tau} &= [-y_2, x_2, p_2] = [x_{02}\sin\tau + y_{02}\cos\tau, x_{02}\cos\tau - y_{02}\sin\tau, p_2] \\ &= [p_1\theta\sin\Sigma - y_1\cos\Sigma, p_2\sin\Sigma \\ &\quad + (x_1 - C)\cos\Sigma, p_2\cos\Sigma - (x_1 - C)\sin\Sigma] \end{aligned} \tag{A.5}$$

The envelope equation, which determines meshing between the surfaces $\mathbf{r}_1$ and $\mathbf{r}_2$:

$$\left(\frac{\partial \mathbf{r}_2}{\partial t} \times \frac{\partial \mathbf{r}_2}{\partial \theta}\right) \cdot \frac{\partial \mathbf{r}_2}{\partial \tau} = 0 \tag{A.6}$$

together with equations for these surfaces, completes a system of equations. If a generating surface 1 is defined by the parameter t, the envelope may be used to calculate another parameter θ, now a function of t, as a meshing condition to define a generated surface 2, now the function of both t and θ. The cross product in the envelope equation represents a surface normal and $\frac{\partial r_2}{\partial \tau}$ is the relative, sliding velocity of two single points on the surfaces 1 and 2 which together form the common tangential point of contact of these two surfaces. Since the equality to zero of a scalar triple product is an invariant property under the applied coordinate system and since the relative velocity may be concurrently represented in both coordinate systems, a convenient form of the meshing condition is defined as:

$$\left(\frac{\partial \mathbf{r}_1}{\partial t} \times \frac{\partial \mathbf{r}_1}{\partial \theta}\right) \cdot \frac{\partial \mathbf{r}_1}{\partial \tau} = -\left(\frac{\partial \mathbf{r}_1}{\partial t} \times \frac{\partial \mathbf{r}_1}{\partial \theta}\right) \cdot \frac{\partial \mathbf{r}_2}{\partial \tau} = 0 \tag{A.7}$$

Insertion of previous expressions into the envelope condition gives:

$$\begin{aligned}&[C - x_1 + (p_1 - p_2)\cot\Sigma]\left(x_1\frac{\partial x_1}{\partial t} + y_1\frac{\partial y_1}{\partial t}\right) \\ &\quad + p_1\left[p_1\theta\frac{\partial y_1}{\partial t} + (p_2 - C\cot\Sigma)\frac{\partial x_1}{\partial t}\right] = 0 \end{aligned} \tag{A.8}$$

This is applied here to derive the condition of meshing action for crossed helical gears of uniform lead with nonparallel and nonintersecting axes. The method constitutes a gear generation procedure which is generally applicable. It can be used for synthesis purposes of screw compressor rotors, which are effectively helical gears with parallel axes. Formed tools for rotor manufacturing are crossed helical gears on non parallel and non intersecting axes with a uniform lead, as in the case of hobbing, or with no lead as in formed milling and grinding. Templates for rotor inspection are the same as planar rotor hobs. In all these cases the tool axes do not intersect the rotor axes.

Accordingly the notes present the application of the envelope method to produce a meshing condition for crossed helical gears. The screw rotor gearing is then given as an elementary example of its use while a procedure for forming a hobbing tool is given as a complex case.

The shaft angle Σ, centre distance C, and unit leads of two crossed helical gears, p_1 and p_2 are not interdependent. The meshing of crossed helical gears is still preserved: both gear racks have the same normal cross section profile, and the rack helix angles are related to the shaft angle as $\Sigma = \psi_{r1} + \psi_{r2}$. This is achieved by the implicit shift of the gear racks in the x direction forcing them to adjust accordingly to the appropriate rack helix angles. This certainly includes special cases, like that of gears which may be orientated so that the shaft angle is equal to the sum of the gear helix angles: $\Sigma = \psi_1 + \psi_2$. Furthermore a centre distance may be equal to the sum of the gear pitch radii: $C = r_1 + r_2$.

Pairs of crossed helical gears may be with either both helix angles of the same sign or each of opposite sign, left or right handed, depending on the combination of their lead and shaft angle Σ.

The meshing condition can be solved only by numerical methods. For the given parameter t, the coordinates x_{01} and y_{01} and their derivatives $\frac{\partial x_{01}}{\partial t}$ and $\frac{\partial y_{01}}{\partial t}$ are known. A guessed value of parameter θ is then used to calculate x_1, y_1, $\frac{\partial x_1}{\partial t}$ and $\frac{\partial y_1}{\partial t}$. A revised value of θ is then derived and the procedure repeated until the difference between two consecutive values becomes sufficiently small.

For given transverse coordinates and derivatives of gear 1 profile, θ can be used to calculate the x_1, y_1, and z_1 coordinates of its helicoid surfaces. The gear 2 helicoid surfaces may then be calculated. Coordinate z_2 can then be used to calculate τ and finally, its transverse profile point coordinates x_2, y_2 can be obtained.

A number of cases can be identified from this analysis.

(i) When $\Sigma = 0$, the equation meets the meshing condition of screw machine rotors and also helical gears with parallel axes. For such a case, the gear helix angles have the same value, but opposite sign and the gear ratio $i = p_2/p_1$ is negative. The same equation may also be applied for the generation of a rack formed from gears. Additionally it describes the formed planar hob, front milling tool and the template control instrument.

(ii) If a disc formed milling or grinding tool is considered, it is sufficient to place $p_2 = 0$. This is a singular case when tool free rotation does not affect the meshing process. Therefore, a reverse transformation cannot be obtained directly.

(iii) The full scope of the meshing condition is required for the generation of the profile of a formed hobbing tool. This is therefore the most complicated type of gear which can be generated from it.

B

Reynolds Transport Theorem

Following Hanjalic, 1983, Reynolds Transport Theorem defines a change of variable ϕ in a control volume V limited by area $\mathbf{A}$ of which vector the local normal is $d\mathbf{A}$ and which travels at local speed $\mathbf{v}$. This control volume may, but need not necessarily coincide with an engineering or physical material system. The rate of change of variable ϕ in time within the volume is:

$$\left(\frac{\partial \phi}{\partial t}\right)_V = \frac{\partial}{\partial t}\int\limits_V \rho\phi dV \tag{B.1}$$

Therefore, it may be concluded that the change of variable ϕ in the volume V is caused by:

- change of the specific variable $\varphi = \Phi/m$ in time within the volume because of sources (and sinks) in the volume, $\left(\frac{\partial \varphi}{\partial t}\right) dV$ which is called a local change and
- movement of the control volume which takes a new space with variable φ in it and leaves its old space, causing a change in time of φ for $\rho\varphi\mathbf{v}.d\boldsymbol{A}$ and which is called convective change.

The first contribution may be represented by a volume integral:

$$\int\limits_V \frac{\partial\left(\rho\varphi\right)}{\partial t} dV \tag{B.2}$$

while the second contribution may be represented by a surface integral:

$$\int\limits_A \rho\varphi\,\mathbf{v}\cdot d\mathbf{A} \tag{B.3}$$

Therefore:

$$\left(\frac{\partial \phi}{\partial t}\right)_V = \frac{d}{dt}\int\limits_V \rho\varphi dV = \int\limits_V \frac{\partial(\rho\varphi)}{\partial t} dV + \int\limits_A \rho\varphi\,\mathbf{v}\cdot d\mathbf{A} \tag{B.4}$$

which is a mathematical representation of Reynolds Transport Theorem.

Applied to a material system contained within the control volume V_m which has surface A_m and velocity $\mathbf{v}$ which is identical to the fluid velocity $\mathbf{w}$, Reynolds Transport Theorem reads:

$$\left(\frac{\partial \phi}{\partial t}\right)_{Vm} = \frac{d}{dt}\int\limits_{Vm} \rho\varphi dV = \int\limits_{Vm} \frac{\partial(\rho\varphi)}{\partial t} dV + \int\limits_{Am} \rho\varphi\,\mathbf{w}\cdot d\mathbf{A} \tag{B.5}$$

If that control volume is chosen at one instant to coincide with the control volume V, the volume integrals are identical for V and V_m and the surface integrals are identical for A and A_m, however, the time derivatives of these integrals are different, because the control volumes will not coincide in the next time interval. However, there is a term which is identical for the both times intervals:

$$\int\limits_V \frac{\partial(\rho\varphi)}{\partial t} dV = \int\limits_{Vm} \frac{\partial(\rho\varphi)}{\partial t} dV \tag{B.6}$$

therefore,

$$\left(\frac{\partial \phi}{\partial t}\right)_{Vm} - \int\limits_{Am} \rho\varphi\,\mathbf{w}\cdot d\mathbf{A} = \left(\frac{\partial \phi}{\partial t}\right)_V - \int\limits_A \rho\varphi\,\mathbf{v}\cdot d\mathbf{A} \tag{B.7}$$

or:

$$\left(\frac{\partial \phi}{\partial t}\right)_{Vm} = \left(\frac{\partial \phi}{\partial t}\right)_V + \int\limits_A \rho\varphi\,(\mathbf{w}-\mathbf{v})\cdot d\mathbf{A} \tag{B.8}$$

If the control volume is fixed in the coordinate system, i.e. if it does not move, $\mathbf{v} = 0$ and consequently:

$$\left(\frac{\partial \phi}{\partial t}\right)_V = \int\limits_V \frac{\partial(\rho\varphi)}{\partial t} dV \tag{B.9}$$

therefore:

$$\left(\frac{\partial \phi}{\partial t}\right)_{Vm} = \int\limits_V \frac{\partial(\rho\varphi)}{\partial t} dV + \int\limits_A \rho\varphi\,\mathbf{w}\cdot d\mathbf{A}\,. \tag{B.10}$$

Finally application of Gauss theorem leads to the common form:

$$\left(\frac{\partial \phi}{\partial t}\right)_{Vm} = \int\limits_V \frac{\partial(\rho\varphi)}{\partial t} dV + \int\limits_V \nabla\cdot(\rho\varphi\,\mathbf{w})\,dV \tag{B.11}$$

As stated before, a change of variable ϕ is caused by the sources q within the volume V and influences outside the volume. These effects may be proportional to the system mass or volume or they may act at the system surface.

The first effect is given by a volume integral and the second effect is given by a surface integral.

$$\left(\frac{\partial\phi}{\partial t}\right)_{Vm} = \int\limits_{Vm} q_V dV + \int\limits_{Am} q_A \cdot d\mathbf{A} = \int\limits_{V} (q_V + \nabla \cdot q_A)\, dV = \int\limits_{V} q dV \quad \text{(B.12)}$$

q can be scalar, vector or tensor.

The combination of the two last equations gives:

$$\int\limits_{V} \frac{\partial(\rho\varphi)}{\partial t} dV + \int\limits_{A} \rho\varphi\, \mathbf{w} \cdot d\mathbf{A} = \int\limits_{V} q dV \quad \text{or}$$

$$\int\limits_{V} \left[\frac{\partial(\rho\varphi)}{\partial t} + \nabla \cdot (\rho\varphi\, \mathbf{w}) - q\right] dV = 0 \quad \text{(B.13)}$$

Omitting integral signs gives:

$$\frac{\partial(\rho\varphi)}{\partial t} + \nabla \cdot (\rho\varphi\, \mathbf{w}) - q = 0 \quad \text{(B.14)}$$

This is the well known conservation law form of variable $\phi = \rho\varphi$. Since for $\varphi = 1$, this becomes the continuity equation: $\frac{\partial\rho}{\partial t} + (\nabla \cdot \rho\mathbf{w}) = 0$ finally it is:

$$\varphi\left[\frac{\partial\rho}{\partial t} + (\nabla \cdot \rho\mathbf{w})\right] + \rho\frac{\partial\varphi}{\partial t} + \rho\,(\mathbf{w} \cdot \nabla\varphi) - q = 0 \quad \text{or}$$

$$\frac{D\varphi}{dt} = \rho\frac{\partial\varphi}{\partial t} + \rho\,(\mathbf{w} \cdot \nabla\varphi) = q \quad \text{(B.15)}$$

$D\varphi/dt$ is the material or substantial derivative of variable φ. This equation is very convenient for the derivation of particular conservation laws. As previously mentioned $\varphi = 1$ leads to the continuity equation, $\varphi = \mathbf{u}$ to the momentum equation, $\varphi = e$, where e is specific internal energy, leads to the energy equation, $\varphi = s$, to the entropy equation and so on.

If the surfaces, where the fluid carrying variable Φ enters or leaves the control volume, can be identified, a convective change may conveniently be written:

$$\int\limits_{A} \varphi\, \rho\mathbf{w} \cdot d\boldsymbol{A} = \int \varphi\, d\dot{m} = (\bar{\varphi}\dot{m})_{\text{in}} - (\bar{\varphi}\dot{m})_{\text{out}} = \dot{\Phi}_{\text{in}} - \dot{\Phi}_{\text{out}} \quad \text{(B.16)}$$

where the overscores indicate the variable average at entry/exit surface sections. This leads to the macroscopic form of the conservation law:

$$\left(\frac{d\Phi}{dt}\right)_{V} = \left[\frac{d(\rho\varphi)}{dt}\right]_{V} = \dot{\Phi}_{\text{in}} - \dot{\Phi}_{\text{out}} + Q = (\bar{\varphi}\dot{m})_{\text{in}} - (\bar{\varphi}\dot{m})_{\text{out}} + Q \quad \text{(B.17)}$$

which states in words: (rate of change of Φ) = (inflow Φ) − (outflow Φ) + (source of Φ)

C

Estimation of Working Fluid Properties

Thermodynamic properties of pure fluids can be obtained by the use of equations of state contained in software packages. Examples of these are the IIR (International Institution of Refrigeration) routines, the THERPROP property package, developed by the authors and, more recently, those of NIST (National Institute of Standards). The NIST routines, which are being progressively updated, are becoming increasingly powerful as they can also be applied to estimate the properties of mixtures.

The equations of state used in these routines vary. Generally, those based on developments of the original Benedict, Webb, Rubin (BWR) 1940,1942 are the most accurate for obtaining pressure-volume-temperature relationships for non-polar fluids and these have the advantage of being equally applicable to both liquid and vapour phases. To obtain useful thermodynamic properties from them, they must be combined with a vapour pressure equation, such as that of Cox-Antoine and an equation for ideal gas specific heats, usually in the form of a 3rd or 4th order polynomial with absolute temperature as the variable. Generalised thermodynamic relationships are then applied to the equation of state used, in order to obtain real gas corrections to specific enthalpy and entropy values obtained from the ideal gas condition.

The BWR type equations, which are explicit in terms of temperature and volume, from which pressure is the derivative, are complex and involve 5th order polynomial terms. They therefore require iterative solution to meet the standard engineering requirement of obtaining thermodynamic properties, given temperature and pressure as the measured input variables. In the case of fluid mixtures, this takes up too much computational time. It is therefore preferable to use simpler cubic equations of state, which can be solved algebraically, without numerical iteration. Iteration is, however, then still needed to obtain the balance between the various mixture components.

The following procedure, used by the Smith and Pitanga, 1994, to obtain the properties of hydrocarbon binary mixtures, is given to illustrate the main principles involved.

Thermodynamic Relationship

The Redlich-Kwong-Soave Equation of State The Redlich-Kwong-Soave, Soave, 1972 (RKS) equation of state can be written as:

$$p = \frac{RT}{V-b} - \frac{a(T)}{V(V+b)} \tag{C.1}$$

Substituting:

$$\frac{ZRT}{p} = V, \quad A = \frac{ap}{R^2T^2} \quad \text{and} \quad B = \frac{bp}{RT} \tag{C.2}$$

This can be rewritten in terms of the compressibility factor Z as:

$$Z^3 - Z^2 + \left(A - B - B^2\right) Z - AB = 0 \tag{C.3}$$

In the case of a mixture of fluids, for each component i at its critical point the intermolecular parameter a and the co-volume b are expressed as follows:

$$a_i(T_{c_i}) = \frac{\Omega_a R^2 T_{c_i}^2}{p_{c_i}} \quad \text{and} \quad b_i(T_{c_i}) = \frac{\Omega_b R T_{c_i}}{p_{c_i}} \tag{C.4}$$

Where the values of the constants Ω_a and Ω_b are given as:

$$\Omega_a = \frac{1}{9(2^{\frac{1}{3}} - 1)} \quad \text{and} \quad \Omega_b = \frac{(2^{\frac{1}{3}} - 1)}{3} \tag{C.5}$$

Soave noted that a successful correlation of phase equilibria of mixtures implied the correlation of the vapour pressures of pure substances. He defined a dimensionless function $\alpha(T_r)$ as follows:

$$\alpha(T_r) = \frac{a(T)}{a(T_c)} \quad \text{where} \quad \alpha(T_r) \to 1 \quad \text{as} \quad T \to T_c \tag{C.6}$$

For non-polar and slightly polar fluids, this function can be expressed in a linear form as:

$$\sqrt{\alpha} = 1 + m\left(1 - \sqrt{T_r}\right) \tag{C.7}$$

where m is a function of the acentric factor ω.

This function was originally defined by Soave, Redlich and Kwong 1949, but later correlated more accurately by Graboski and Daubert, 1955 as:

$$m = 0.48508 + 1.55171\,\omega_i \; - \; 0.15613\,\omega_i^2 \tag{C.8}$$

The acentric factor was originally defined by Pitzer et al., 1955 and 1957. The more recent definition of Lee and Kesler, 1975 has wide usage and is given as:

$$\omega = \frac{\ln p_{br} - 5.92714 + \frac{6.09648}{T_{br}} + 1.28862 \ln T_{br} - 0.169347 T_{br}^6}{15.2518 - \frac{15.6875}{T_{br}} - 13.4721 \ln T_{br} + 0.43577 T_{br}^6} \quad \text{(C.9)}$$

Soave's method includes classical mixing rules for the determination of vapour-liquid equilibrium parameters when mixtures are non-polar and weakly polar. These contain an adjustable binary interaction parameter (k_{ij}):

$$a = \sum_{i=1}^{n} \sum_{j=1}^{n} x_i x_j a_{ij} \quad \text{(C.10)}$$

where:

$$a_{ij} = \sqrt{a_i a_j}(1 - k_{ij}) \quad \text{(C.11)}$$

$$b = \sum_{i=1}^{n} x_i b_i \quad \text{(C.12)}$$

Vapour-Liquid Equilibrium Calculation

Vapour-liquid equilibrium implies the equality of fugacities.

Hence:

$$f_i^l = f_i^v (i = 1, \dots n) \quad \text{(C.13)}$$

The fugacities are normally replaced by the fugacity coefficient since it is more directly related to the measurable properties pressure, temperature and mole fraction.

$$\phi_i^l (T, p, x_i) \, x_i = \phi_i^v (T, p, y_i) \, y_i \quad \text{(C.14)}$$

Fugacity coefficients are derived analytically and for Soave's equation of state can be expressed as follows:

$$\ln \phi_i = \frac{b_i}{b}(z - 1) - \ln(z - B) + \frac{A}{B}\left[\frac{b_i}{b} - \frac{2}{a}\sum_j y_j a_{ij}\right] \ln\left(1 + \frac{B}{z}\right) \quad \text{(C.15)}$$

Thermodynamic Property Estimation

The equations used for evaluating thermodynamic properties were as follows:

Enthalpy

$$H = H^0 - \Delta H + H_{\text{Correc}} \quad \text{(C.16)}$$

The first term on the right hand side of (2.4) is obtained from the ideal gas heat capacities of the pure components. These are a function of temperature and are normally expressed in a polynomial form as:

$$C_p^0 = A + BT + CT^2 + DT^3 + \cdots \tag{C.17}$$

where:

$$A = \sum_i x_i A_i \qquad B = \sum_i x_i B_i \quad \text{etc.} \tag{C.18}$$

so that:

$$H^0 = \int_{T_0}^{T} C_p^0 dT \tag{C.19}$$

Substituting for the ideal gas heat capacity, it becomes:

$$H^0 = A(T - T_0) + \frac{B}{2}\left(T^2 - T_0^2\right) + \frac{C}{3}\left(T^3 - T_0^3\right) + \frac{D}{4}\left(T^4 - T_0^4\right) \tag{C.20}$$

The values for A, B, C and D were obtained from the THERPROP databank.

ΔH, the enthalpy departure function, for the RKS equation is given as:

$$\Delta H = RT\left[(1 - z) + \frac{A}{B}\left(1 + \frac{D}{a}\right) . \ln\left(1 + \frac{B}{z}\right)\right] \tag{C.21}$$

where:

$$D = \sum_i \sum_j y_i y_j m_j \left(1 - k_{ij}\right) \sqrt{a_i \alpha_i} \sqrt{a_j Tr_j} \tag{C.22}$$

H_{Correc} sets a common base value ($h_f = 100\,\mathrm{kJ/kg}$ at 0°C)

Entropy

$$S = S^0 - \Delta S + S_{\mathrm{Correc}} + S_{\mathrm{Mix}} \tag{C.23}$$

where:

$$S^0 = A \ln\left(\frac{T}{T_0}\right) + B(T - T_0) + \frac{C}{2}\left(T^2 - T_0^2\right) + \frac{D}{3}\left(T^3 - T_0^3\right) + R \ln\left(\frac{p}{p_0}\right) \tag{C.24}$$

ΔS, is the entropy departure function, corrected from the given value as, Walas, 1985:

$$\Delta S = -R \ln(z - B) + \frac{AD}{Ba} \ln\left(1 + \frac{B}{z}\right) \tag{C.25}$$

S_{Mix}, the entropy of mixing is given by:

$$S_{\mathrm{Mix}} = -R y_i \ln y_i \tag{C.26}$$

and S_{Correc} sets a common base value ($s_f = 1.0\,\mathrm{kJ/kg}$ at 0°C)

Computational Techniques

Equations of state can be represented graphically on pressure-volume coordinates as a family of isotherms, as shown in Fig C.1. In the case of fluid mixtures, unlike pure fluids, dew and bubble temperatures at the same vapour pressure are not normally equal. The mode of estimation of saturation conditions may be shown most simply when the temperature is given and the corresponding pressure is required. The equation is then reduced to the case of a single isotherm. A cubic equation, such as the RKS, yields three roots in the two phase region. the largest of these corresponds to the vapour phase volume, the smallest to that of the liquid phase while the intermediate root, which is located on a positive gradient, has no physical significance. The pressure is iterated until equal fugacity values are obtained from the roots corresponding to liquid and vapour.

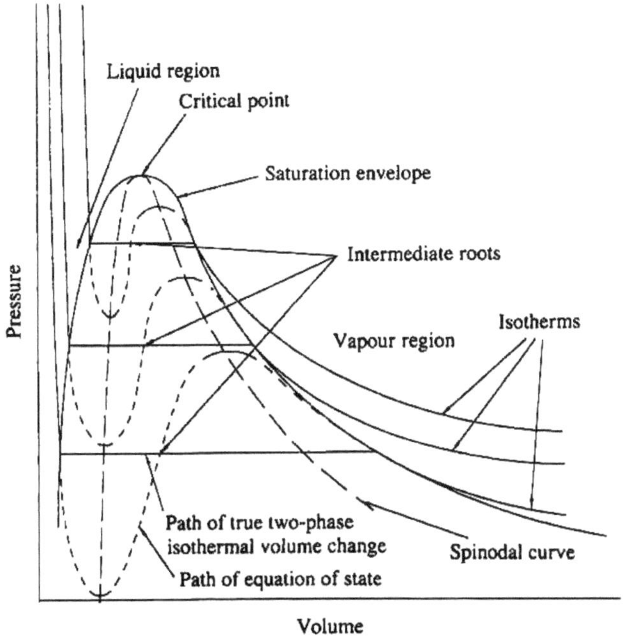

Fig. C.1. p-v diagram for a fluid in the two-phase region with isotherms derived from a cubic equation of state

During the process of iteration, it is possible that values of pressure may be derived for which only one real root exists, as in the pure liquid and dry vapour regions shown in Fig. 2.3. In such cases, in order to enable the solution to continue until the required two roots are attained the method of solution must contain procedures:

(i) To determine whether the single root is a liquid or vapour value.
(ii) To assign a hypothetical real root to the other phase for the fugacity comparison.

Various techniques have been used for this purpose, for example, Poling et al., 1981; Gundersen, 1982 and Ziervogel and Poling, 1983. The following method was used in this investigation.

The locus of the maxima and minima of all the subcritical isotherms was obtained. For all cubic equations of state it is known as the van der Waals spinodal curve and for the RKS equation it has the form:

$$V^4 + 2\left[b - \frac{a}{RT}\right]V^3 + \left[b^2 + \frac{3ab}{RT}\right]V^2 - \frac{ab^3}{RT} = 0 \qquad \text{(C.27)}$$

The roots of this were established for the isotherm under consideration. These are fixed in bubble and dew pressure calculations but vary with each iteration in bubble and dew temperature estimation procedures. The van der Waals spinodal is a fourth order polynomial in volume. Hence it has four roots. Imaginary and negative roots, which have no thermodynamic significance, were discarded. Since the repulsive term in any cubic equation leads to a singularity at $V = b$ it was found that only roots greater than co-volume b could be used. The single root of the RKS equation was then compared with the remaining spinodal roots. If it was larger than the biggest spinodal root it was assumed to be a vapour root. The smallest spinodal root was taken as the liquid root. If the RKS root was smaller than the smallest spinodal root it was assumed to be a liquid root and the largest spinodal root taken as the vapour value.

Other problems include the selection of suitable starting values for the iterative procedure. At low pressures, the assumption of Raoult's law was sufficient for a first estimate but at higher pressures, this did not lead to convergence. A variety of checks and modifications therefore had to be applied to the starting value before iteration began, in order to obtain equal fugacities. The method of iteration then used was the Newton-Raphson procedure.

References

Amosov P.E. et al., 1977: Vintovie kompresornie mashinii – Spravochnik (Screw Compression Machines-Handbook), Mashinstroienie, Leningrad

Andreev P.A., 1961: Vintovie kompressornie mashinii (Screw Compression Machines), SUDPROM Leninngrad

Arbon I.M., 1994: The Design and Application of Rotary Twin-shaft Compressors in the Oil and Gas Process Industry, MEP London

Astberg A., 1982: Patent GB 2092676B

Bammert K., 1979: Patent Application FRG 2911415

Benedict M., Webb G.B. and Rubin L.C., 1940: An empirical equation for thermodynamic properties of light hydrocarbons and their mixtures. I Methane, ethane, propane and butane, *Jrnl Chem Physics*, V. 8 pp. 334–345

Benedict M., Webb G.B. and Rubin L.C., 1942: An empirical equation for thermodynamic properties of light hydrocarbons and their mixtures. II Mixtures of methane, ethane, propane and n-butane, *Jrnl Chem Physics* V. 10 pp. 747–758

Benedict M., Webb G.B. and Rubin L.C., 1951: An empirical equation for thermodynamic properties of light hydrocarbons and their mixtures. Fugacities and liquid-vapour equilibria, *Chem Eng Progress*, V. 47, n8, pp. 419–422, n9 pp. 449–454, 571–578 and 609–620

Bowman J.L., 1983: US Patent 4,412,796

Box, M.J., 1965: A new method of constrained optimisation and a comparison with other methods, Computer Journal, V. 8, pp. 42–52

Brasz J.J., Shistla V., Stosic N. and Smith, I.K., 2000: Development of a Twin Screw Expressor as a Throttle Valve Replacement for Water-Cooled Chillers, *XV International Compressor Engineering Conference* at Purdue, July 2000

Buckingham E., 1963: Analytical Mechanics of Gears, Dover Publ, New York

Chia-Hsing C., 1995: US Patent 5,454,701

Colbourne J.R., 1987: The Geometry of Involute Gears, Springer Verlag, New York

Demirdzic I., Peric M., 1990: Finite Volume Method for Prediction of Fluid flow in Arbitrary Shaped Domains with Moving Boundaries, *Int. J. Numerical Methods in Fluids* V. 10, 771

Demirdzic I., Muzaferija S., 1995: Numerical Method for Coupled Fluid Flow, Heat Transfer and Stress Analysis Using Unstructured Moving Mesh with Cells of Arbitrary Topology, *Comp. Methods Appl. Mech Eng*, V. 125 235–255

Edstroem S.E., 1974: US Patent 3,787,154

Edstroem S.E., 1989: Quality Classes for Screw Compressor Rotors, *Proceedings of IMechE Conference Development in Industrial Compressors*, 83

Edstroem S.E., 1992: A Modern Way to Good Screw Compressors, *International Compressor Engineering Conference* At Purdue, 18

Ferziger J.H., Peric M., 1996: Computational Methods for Fluid Dynamics, Springer, Berlin

Fleming J.S., Tang Y., 1994: The Analysis of Leakage in a Twin Screw Compressor and its Application to Performance Improvement, Proceedings of IMechE, *Journal of Process Mechanical Engineering*, V. 209, 125

Fleming J.S., Tang Y. and Cook G., 1998: The Twin Helical Screw Compressor, Part 1: Development, Applications and Competetive Position, Part 2: A Mathematical Model of the Working process, Proceedings of the IMechE, *Journal of Mechanical Engineering Science*, V. 212, p. 369

Fujiwara M., Osada Y., 1995: Performance Analysis of Oil Injected Screw Compressors and their Application, *Int J Refrig* V. 18, 4

Golovintsov A.G. et al., 1964: Rotatsionii kompresorii (Rotary Compressors), Mashinostroenie, Moscow

Graboski M.S. and Daubert T.E., 1955: Modified Soave equation of state for phase equilibrium calculations, *Ind Eng Chem Process Des Div*, V. 18, n2, pp. 300–306

Gundersen T., 1982: Numerical aspects of the implementation of cubic equations of state in flash calculation routines, *Comp Chem Eng*, V. 6, n3, pp. 245–255

Hanjalic K., 1983: Gas Dynamics, Svjetlost Sarajevo, 1983

Hanjalic K., Stosic N., 1994: Application of mathematical modeling of screw engines to the optimization of lobe profiles, Proc. VDI Tagung "Schraubenmaschinen 94" Dortmund VDI Berichte Nr. 1135

Hanjalic K., Stosic N., 1997: Development and Optimization of Screw Machines with a Simulation Model, Part II: Thermodynamic Performance Simulation and Design, ASME Transactions, *Journal of Fluids Engineering*, V. 119, p. 664

Holmes C.S., 1994: Towards a Core Program for the Measurement of Screw Rotor Bodies by Co-ordinate Measuring Machine, Proc. VDI Tagung "Schraubenmaschinen 94", Dortmund VDI Berichte 1135

Holmes C.S., Stephen A.C., 1999: Flexible Profile Grinding of Screw Compressor Rotors, International Conference on Compressors and Their Systems, IMechE London

Hough D., Morris S.J., 1984: Patent Application GB 8413619
Kasuya K. et al., 1983: US Patent 4,406,602
Kauder K., Harling H.B., 1994: Visualisierung der Oelverteilungin Schraubenkompressoren (Visualisation of Oil Effects in ScrewCompressors), Proc. VDI Tagung "Schraubenmaschinen 94", Dortmund VDI Berichte 1135
Kauder K., Helpertz M., 1998: Einlauf- und Hybridschichten fuer Schraubenkompressoren (Run-in and Hybrid Coatings for Twin-Screw Compressors), Proc. VDI Tagung "Schraubenmaschinen 98", Dortmund VDI Berichte 1391
Konka K-H., 1988: Schraubenkompressoren (ScrewCompressors) VDI-Verlag, Duesseldorf
Kovacevic A., Stosic N. and Smith, I.K., 2003: Three Dimensional Numerical Analysis of Screw Compressor Performance, Journal of Computer Methods in Applied Mechanics and Engineering, V. 3, n2, pp. 259–284
Kovacevic A., Stosic N, Smith, I.K. and Mujic E., 2004: Development of a Management Interface for Screw Compressor Design Tools, Proceedings of the TMCE 2004, April 13–17, Lausanne
Lee B.I., Kesler M.G., 1975: A generalised thermodynamic correlation based on Pitzer's three parameter corresponding states principle, *AIChE Jrnl*, V. 21, n3 pp. 510–527
Lee H-T., 1988: US Patent 4,890,992
Litvin F.L., 1956: Teoria zubchatiih zaceplenii (Theory of Gearing), Nauka Moscow, second edition 1968, also Gear Geometry and Applied Theory Prentice-Hill, Englewood Cliffs, NJ 1994
Lysholm A., 1942: A New Rotary Compressor, *Proc.* IMechE 150,11
Lysholm A., 1966: The Fundamentals of a New Screw Engine, ASME Paper No 66-GT-86
Lysholm A., 1967: US Patent 3,314,598
Margolis D.L., 1978: Analytical Modelling of Helical Screw Turbines for Performance Prediction, *J. Engr. for Power* 100(3)482
McCreath P., Stosic N., Smith I.K. and Kovacevic A., 2001: The Design of Efficient Screw Compressors for Delivery of Dry Air, International Conference on Compressors and Their Systems, IMechE, London
Menssen E., 1977: US Patent 4,028,026
Meyers K., 1997: Creating the Right Environment for Compressor Bearings, Evolution, *SKF Industrial Journal*, V. 4, 21
O'Neill P.A., 1993: Industrial Compressors, Theory and Equipment, Butterworth-Heinemann, Oxford
Nilson, 1952: US Patent 2,622,787
Ohman H., 1999: PCT Patent WO 98/27340
Peng N., Xing Z., 1990: New Rotor Profile and its Performance Prediction of Screw Compressor, International Compressor Engineering Conference At Purdue, 18

Pitzer K.S., Lippmann D.Z., Curl jr R.F., Huggins C.M. and Petersen D.E., 1955: The volumetric and thermodynamic properties of fluids. I Theoretical basis and virial coefficients, *Jrnl Am Chem Soc*, V. 77, n13, pp. 3427–3432

Pitzer K.S., Lippmann D.Z., Curl jr R.F., Huggins C.M. and Petersen D.E., 1955: The volumetric and thermodynamic properties of fluids. II Compressibility factor, vapor pressure and entropy of vaporisation, *Jrnl Am Chem Soc*, V. 77, n13, pp. 3433–3440

Pitzer K.S., Curl jr R.F., 1957: The volumetric and thermodynamic properties of fluids. III Empirical equation for the second virial coefficient, *Jrnl Am Chem Soc*, V. 79, p. 2369

Poling B.E., Grens E.A. and Prausnitz J.M., 1981: Thermodynamic properties from a cubic equation of state – avoiding trivial roots and spurious derivatives, *Ind Eng Chem Process Des Div*, V. 20, n1, pp. 127–130

Redlich O., Kwong J.N.S., 1949: On the thermodynamics of solutions. V. An equation of state. *Fugacities of gaseous solutions, Chem Rev* V. 44, pp. 233–244

Rinder L., 1979: Schraubenverdichter (Screw Compressors), Springer Verlag, New York

Rinder L., 1984: Schraubenverdichterlaeufer mit Evolventenflanken (Screw Compressor Rotor with Involute Lobes), Proc. VDI Tagung "Schraubenmaschinen 84" VDI Berichte Nr. 521 Duesseldorf

Rinder L., 1987: US Patent 4,643,654

Sauls J., 1994: The Influence of Leakage on the Performance of Refrigerant Screw Compressors, Proc. VDI Tagung "Schraubenmaschinen 94", Dortmund VDI Berichte 1135

Sauls J., 1998: An Analytical Study of the Effects of Manufacturing on Screw Rotor Profiles and Rotor Pair Clearances, Proc. VDI Tagung "Schraubenmaschinen 98", Dortmund VDI Berichte 1391

Sakun I.A., 1960: Vintovie kompresorii (Screw Compressors), Mashinostroenie Leningrad 41.

Shaw D.N., 1999 Patent US 5,911,734

Shibbie, 1979: US Patent 4,140,445

Singh P.J., Onuschak A.D., 1984: A Comprehensive Computerized Method For Twin Screw Rotor Profile Generation and Analysis, Purdue Compressor Technology Conference 544

Singh P.J., Schwartz J.R., 1990: Exact Analytical Representation of Screw Compressor Rotor Geometry, International Compressor Engineering Conference At Purdue, 925

Smith I.K., Pitanga Marques da Silva R., 1994: Development of the trilateral flash cycle system, Part 2: Increasing power output with working fluid mixtures, Proceedings of IMechE, *Journal of Power and Energy*, V. 208, p. 138

Smith I.K., Stosic N. and Aldis C.A., 1996: Development of the trilateral flash cycle system, Part 3: The design of high efficiency two-phase screw expanders, Proceedings of IMechE, *Journal of Power and Energy*, V. 210, p. 75

Soave G., 1972: Equilibrium constants from a modified Redlich-Kwong equation of state, *Chem Eng Science*, V. 27, n6, pp. 1197–1203

Stosic N., Hanjalic K., 1977: Contribution towards Modelling of Two-Stage Reciprocating Compressors, *Int. J. Mech. Sci.* 19, 439

Stosic N., Hanjalic K., Koprivica J., Lovren N. and Ivanovic M., 1986: CAD of Screw Compressor Elements, Strojarstvo Journal Zagreb 28, 181

Stosic N., Kovacevic A., Hanjalic K. and Milutinovic Lj., 1988: Mathematical Modelling of the Oil Influence upon the Working Cycle of Screw Compressors, International Compressor Engineering Conference at Purdue, 355

Stosic N., Milutinovic Lj., Hanjalic K. and Kovacevic A., 1992: Investigation of the Influence of Oil Injection upon the Screw Compressor Working Process, *Int. J. Refrig.* 15,4,206

Stosic N., Hanjalic K., 1994: Development and optimization of screw engine rotor pairs on the basis of computer modeling, Proc. XVII Conference on Compressor Engineering at Purdue, 55

Stosic N., 1996: Patent Application GB 9610289.2

Stosic N., Hanjalic K., 1997: Development and Optimization of Screw Machines with a Simulation Model, Part I: Profile Generation, ASME Transactions, *Journal of Fluids Engineering*, V. 119, p. 659

Stosic N., Smith I.K., Kovacevic A., Aldis C.A., 1997: The Design of a Twin-screw Compressor Based on a New Profile, *Journal of Engineering Design*, V. 8, 389

Stosic N., Kovacevic A. and Smith I.K., 1998: Modelling of Screw Compressor Capacity Control, Proc. XIV Conference on Compressor Engineering at Purdue, 607

Stosic N. et al., 1998: The Performance of a Screw Compressor with Involute Contact Rotors in a Low Viscosity Gas-Liquid Mixture Environment, Proc. VDI Tagung "Schraubenmaschinen 98", Dortmund VDI Berichte 1391

Stosic N., 1998: On Gearing of Helical Screw Compressor Rotors, Proceedings of IMechE, *Journal of Mechanical Engineering Science*, V. 212, 587

Stosic N., 1999: Recent Developments in Screw Compressors, International Conference on Compressors and Their Systems, IMechE London

Stosic N., Smith I.K., Kovacevic A. and Venu Madhav K., 2000: Retrofit "N" Rotors for Efficient Oil-Flooded Screw Compressors, International Conference on Compressors and Their Systems, ImechE London

Stosic N., Smith I.K. and Kovacevic A., 2001: Calculation of Rotor Interference in Screw Compressors, International Compressor Technique Conference, Wuxi, China

Stosic N., Smith I.K., and Kovacevic A., 2002: A Twin Screw Combined Compressor and Expander for CO_2 Refrigeration Systems, XVI International Compressor Engineering Conference at Purdue, July 2002

Stosic N., Smith I.K. and Kovacevic A., 2003a: Rotor Interference as a Criterion for Screw Compressor Design, *Journal of Engineering Design*, 14(2)209

Stosic N., Smith I.K. and Kovacevic A., 2003b: Optimisation of Screw Compressors, *Journal of Applied Thermal Engineering*, 23(10)1177

Stosic N., Smith I.K. and Kovacevic A., 2003c: Opportunities for Innovation with Screw Compressors, Proceedings of IMechE, Part E, *Journal of Process Mechanical Engineering*, 217(3)157

Stosic N., 2004: Screw Compresors in Refrigeration and Air Conditioning, *Int Journal of HVACR Research*, 10(3) pp. 233–263, July 2004

Tang Y., Fleming J.S., 1992: Obtaining the Optimum Geometrical Parameters of a Refrigeration Helical Screw Compressor, International Compressor Engineering Conference at Purdue 213

Tang Y., Fleming J.S., 1994: Clearances between the Rotors of Helical Screw Compressors: Their determination, Optimization and Thermodynamic Consequences, Proceedings of IMechE, *Journal of Process Mechanical Engineering*, V. 208, 155

Tang Y, 1995: Computer Aided Design of Twin Screw Compressors, PhD Thesis, Strathclyde University, Glasgow, UK

Varadaraj J., 2002: GB Patent Application 0211807.3

Venu Madhav K., Stosic N., Smith I.K. and Kovacevic A., 2001: The Design of a Family of Screw Compressors for Oil–Flooded Operation International Conference on Compressors and Their Systems, ImechE London

Walas S.M., 1985: Phase Equilibria in Chemical Engineering, Butterworths, Stoneham, MA, USA

Xing Z.W., 2000a: Screw Compressors: Theory, Design and Application, (in Chinese), China Machine Press, Beijing, China

Xing Z.W., Peng, X. and Shu P., 2000b: Development and Application of a Software Package for Design of Twin Screw Compressors, 2000 International Compressor Conference at Purdue, 1011

Xion Z., Dagang X., 1986: Study on Actual Profile Surface and Engaging Clearance of Screw Compressor Rotors, Purdue Compressor Technology Conference 239

Zenan X., 1984: The Dynamic Measurement and Mating Design of a Screw Compressor Rotor Pair, Purdue Compressor Technology Conference 314

Zhang L., Hamilton J.F., 1992: Main Geometric Characteristics of the Twin Screw Compressor, International Compressor Engineering Conference at Purdue, 213

Zhong J., 2002: US Patent 6,422,846

Ziervogel R.G., Poling B.E., 1983: A simple method for constructing phase envelopes for multicomponent mixtures, *Fluid Phase Equilibria*, V. 11, n2, pp. 127–135

MIX
Papier aus verantwortungsvollen Quellen
Paper from responsible sources
FSC® C105338

If you have any concerns about our products,
you can contact us on
ProductSafety@springernature.com

In case Publisher is established outside the EU,
the EU authorized representative is:
Springer Nature Customer Service Center GmbH
Europaplatz 3, 69115 Heidelberg, Germany

Printed by Libri Plureos GmbH
in Hamburg, Germany